Successful Construction Project Management

The Practical Guide

Paul Netscher

Copyright Note
Copyright © 2014 Paul Netscher
All rights reserved. No part of this publication may be reproduced or transmitted, in whole or in part, by any means without written permission from the publisher.
Published by Panet Publications
PO Box 2119, Subiaco, 6904, Australia

www.pn-projectmanagement.com

ISBN: 978-1497344419

Available from Amazon.com and other retail outlets

Legal Notices

It should be noted that construction projects are varied, use different contracts, abide by different restrictions, regulations, codes and laws, which vary between countries, states, districts and cities. Furthermore various industries have their own distinct guidelines, acts and specific protocols which the contractor must comply with. To complicate matters further these laws, acts and restrictions are continually evolving and changing. Even terminologies vary between counties, industries and contracts and may not be the same as those included in this publication. It's therefore important that readers use the information in this publication, taking cognisance of the particular rules that apply to their project.

Each project has its own sets of challenges and no one book can cover all the steps and processes in every project. This publication covers a broad range of projects without being specific to a particular field of work. Some of the author's personal opinions may not be pertinent to certain projects, clients or companies. Readers should undertake further research and reading on the topics particularly relevant to them, even requesting expert advice when required.

Therefore, the author, publisher and distributor assume no responsibility or liability for any loss or damage, of any kind, arising from the purchaser or reader using the information or advice contained herein.

The examples used in the book should not be seen as a criticism of people or companies, but, should rather be viewed as cases which we all can learn from. After all we've all made mistakes. Any perceived slights are unintentional.

Cover layout by Clark Kenyon, www.camppope.com
Cover photographs by Paul Netscher, João Neto, Ian Weir

Preface

Construction covers many activities from; roads, dams reservoirs, pipelines, houses, factories, airports, mines, bridges, harbours, office buildings, tunnels, railways, water treatment plants, power stations, apartment buildings and minerals process plants. Whatever we touch has passed along routes that have been built by someone.

Some of these structures are iconic, leaving an indelible mark on the landscape, such as the Sydney Opera House, the Golden Gate Bridge, St Paul's Cathedral, the Eiffel Tower and the Empire State Building. However, many structures will never be seen by the public eye – they are either underground, on a mine site hundreds of kilometres from the public gaze, hidden behind security or covered from view. Some structures which have served us well have been left to fall derelict or have been demolished to make way for new structures.

Most structures and buildings have taken countless hours to design and build, some even millions of hours. They have required a coordinated effort from people with a diverse array of skills which may include civil, mechanical and electrical engineers, architects, supervisors, plumbers, electricians, carpenters, tilers, painters, steelworkers, operators, form-workers and labourers. These people have generally had to be welded into a coordinated team to deliver the project on time, within budget, to the right quality and without harming people or the environment. To do this a unique type of person has developed, who is given the responsibility of successfully delivering the project – the Project Manager, also referred to as the Site Manager, Construction Manager or Site Agent. Without this person very few projects would be completed.

However, after working in the construction industry for twenty-eight years, on over one hundred projects, it's apparent that many construction projects are poorly managed, resulting in them being completed late, losing money, having poor quality, often ending up in expensive contractual and legal claims, and in the worst case, causing injury to workers.

This results in the industry having a poor image. Much of this is avoidable if the projects are managed by knowledgeable and experienced Project Managers. Unfortunately they aren't always available.

If anything, the situation appears to be worsening. Projects are fast-track, clients are more demanding, red-tape ties up valuable management time, there are fewer skilled trades-people, and many managers don't have the required experience themselves, nor the time, to mentor and train the Project Managers working for them.

This book seeks to pass on my wealth of knowledge and experience in an easy to read, simple format, providing practical and relevant information, including many examples gathered from my projects.

Acknowledgements

The contents of this book are a result of 28 years of experience in the construction industry. Most of that time has been happy, exciting, challenging, exhilarating and satisfying. Contributing to this has been the numerous wonderful people that I've worked with. From the humblest labourer through to the various managers I reported to over the years, and especially the staff that worked with me. You all left your mark on me in some way, you all helped me be successful, and in some small way you each contributed to this book.

A special thanks to Elizabeth who painstakingly coached me through my initial writing attempts. I'm not sure how you worked out what I was trying to say in my first drafts – but you did. You then edited the complete book for me. Thanks for your patience.

Thanks to Dellas, Kerry and Paul who agreed to read through various chapters to verify the content. I don't think they knew what they were letting themselves in for when they so willingly agreed to undertake the task. Your comments were helpful and insightful.

Thank you to Tim for proof reading the complete book in the short time I gave you. Your comments were invaluable.

Thanks to my family who patiently stood by me, offering their assistance, during this process.

Contents at a Glance

Introduction	xv
Chapter 1 - Planning the Project	1
Chapter 2 - Starting the Project	29
Chapter 3 - Scheduling (Programming) the Project	48
Chapter 4 - Delivering the Project	61
Chapter 5 - Safety and Environment	88
Chapter 6 - People	110
Chapter 7 - Plant and Equipment	121
Chapter 8 - Materials	134
Chapter 9 - Quality Control	151
Chapter 10 - Subcontractors	163
Chapter 11 - Financial	187
Chapter 12 - Contractual	204
Chapter 13 - Completing and Closing Out the Project	217
Conclusion	227
Glossary	228

Contents

Introduction	xv
Skills required to manage a construction project	xv
A challenging and changing environment	xvi
Responsibilities of a Project Manager	xvii
Avoiding mistakes	xviii
How this book will help	xix
Chapter 1 - Planning the Project	**1**
Tender documents	3
Scope of works	4
Reading through the contract documents	4
Tender drawings	5
Access to site	6
Client provided services	6
Site conditions and constraints	6
Tender handover	6
Site visit	7
Planning the execution of the work	7
Schedule	8
Subcontract or self-perform	11
Alternative construction methods	12
Alternative designs	13
Formwork	13
Vertical access	14
Manning	14
Staffing	14
Inductions and site access requirements	17
Choosing the plant and equipment	18
Site establishment	18
Site office requirements	19
Worker facilities	20
Provision of utilities and site services	20
Accommodation of workers and staff	21
Ordering materials	22
Safety and environmental preparations	22
Insurances	23
Bonds	24
Permits	26
Notifications	26
Licenses, registrations, certificates and qualifications	26
Quality management plans	26
Client deliverables	27
Project plan	27
Mobilisation check list	28

Risk review and understanding the risks	28
Summary	28
Chapter 2 - Starting the Project	**29**
Facilities	29
Setting up the safety systems	30
Welcome packs	31
Project induction	32
Staff induction and welcome	34
Site hours	34
Accommodation for personnel	35
Transport of personnel	35
Site fencing and security	36
Access to site and access on site	38
Parking areas	39
Diversions, deviations and road closures	39
Stormwater	40
Signage and posters	41
Lighting	42
Stationery	42
Document control and filing systems	43
Preconstruction survey and photographs	44
Accepting handover of work areas	44
Excavation permits, and location and protection of existing services	44
Survey and setting-out	44
The client	45
Establishing relationships	46
Quality	46
Summary	46
Chapter 3 - Scheduling (Programming) the Project	**48**
Introduction	52
Requirements of a schedule	52
Resource levelling	53
Long-lead items	53
Relationship between time and cost	53
Awareness of client and other imposed restrictions	54
The critical path	54
Agreeing the contract schedule	55
Information required schedule	55
S-curves and histograms	56
Monitoring progress against the schedule	57
Communicating the schedule and progress to the project staff	58
Float	58
Variations	59
Approval of a revised contract schedule	59
Special use schedules	59
Summary	60

Chapter 4 - Delivering the Project — 61
- Delegation — 61
- Time management — 62
- Teamwork — 63
- Communication — 64
- Emails — 65
- Letter writing — 66
- Problem solving — 67
- Decision making — 67
- Negotiation — 67
- Client relations — 68
- Manage client expectations — 68
- Day-to-day planning — 69
- Meeting notes — 69
- Computer systems and networks — 69
- File management — 70
- Drawings — 70
- Late information — 71
- Request for information — 71
- Shop drawings — 72
- Contact list — 73
- Daily records and daily reports — 73
- Weekly and monthly reports — 74
- Client site meetings — 74
- Staff meetings — 76
- Reporting procedures — 77
- Photographs — 77
- Diaries — 78
- Potential problems with the client's design — 78
- Assisting the client's Designers with solutions — 79
- Coordination of services — 79
- Labour productivity — 80
- Plant and equipment productivity — 82
- Contractor management visits — 83
- Client management visits — 84
- Company inter-departmental visits and audits — 84
- Weather damage — 85
- Making changes — 85
- Help — 86
- Opportunities for further work and marketing — 86
- Summary — 86

Chapter 5 - Safety and Environment — 88
- What standard should be set for the safety on the project? — 90
- Planning and preparation — 91
- Safety leadership — 91
- Rules — 91
- Documentation — 92

Safety plans 93
Registers 93
Tagging of equipment 93
Toolbox meetings 94
Safety meetings 94
Inspections 95
Safety audits 95
Safety training 95
First-aid facilities 95
In the event of an accident 96
Accident and incident reporting 96
Accident investigation 97
Safety communication 98
Safety signage 98
Safe equipment 98
Traffic 98
Housekeeping 99
Material deliveries, storage and handling 100
Hazardous materials 100
Weather 100
Fatigue management 103
Fire 104
Injury to a member of the public 104
Safety in operational facilities 105
Substance abuse 106
Environmental 106
Waste handling and disposal 107
Fuel, oil and other liquids 108
Summary 108

Chapter 6 - People **110**
Company policies and procedures 111
Employment contracts 111
Personnel records 112
Identity documents, passports, visas, permits and qualifications 113
Discipline 113
Lead by example 113
Drugs and alcohol 114
Friends and relations 114
Indigenous and local employment 114
Cultures, ethnicity and backgrounds 115
Training 115
Feedback 116
Grievance procedures 117
High labour turnover 117
Leave 118
Gifts, bribes and corruption 118
Unions 119

The art of persuasion	119
Conflict resolution	120
Summary	120
Chapter 7 - Plant and Equipment	**121**
The right item for the job	121
Internal hire, external hire or purchase	121
External hire	122
Ordering equipment	123
Transporting the equipment	123
When the equipment arrives on site	125
Understanding the conditions in the hire agreement	126
Booking of plant hours	126
Care and maintenance	127
Service records	128
Calibration records	128
Protection of equipment	128
Overloading of equipment	129
Reporting breakdowns	129
Operator training	129
Formwork and scaffolding	129
Formwork and scaffolding design	130
Cranes, lifting equipment and slings	131
Returning equipment off hire	132
Summary	133
Chapter 8 - Materials	**134**
Ordering materials – quantities	134
Ordering materials – specification	136
Ordering materials – transport	137
Ordering materials – offloading and handling	137
Ordering materials – installation	138
Ordering materials – delivery dates	138
Ordering materials – quality procedures	139
Quotes and tenders	139
Is the cheapest really the cheapest? – (adjudicating quotations)	140
Purchase orders	141
Samples	141
Batches	141
Procuring from a foreign supplier	142
Source of materials – environmental, health, safety and legal considerations	142
Impartial tender process	142
Alternative materials	143
Expediting and managing material deliveries	143
Transport of materials to the project	143
Transport of materials and equipment cross-border	145
Receiving and storing materials	146
Reporting of defective materials	148
Contamination of materials	148

Material handling	148
Supplier's shop drawings	149
Materials supplied by the client	149
On site measurements and templates	149
Summary	150
Chapter 9 - Quality Control	**151**
Responsibility	154
Quality plans	155
Quality documentation	155
Non-conformance reports	156
Quality of materials	157
Reporting and tracking defective material	158
Training to achieve better quality	159
Subcontractor quality	159
Testing and inspection	159
Punch lists	160
Recognition	161
Summary	161
Chapter 10 - Subcontractors	**163**
Types of subcontractors	165
Subcontract tenders	166
Selecting a subcontractor	166
Bid shopping	168
Labour and equipment supply subcontract documentation	168
Adjudicating subcontract tenders	169
Subcontract negotiation	169
Subcontractor documentation	170
Preconstruction (or kick-off) meetings	170
Communication with project staff	171
Subcontractor safety	171
Approval of staff and equipment	172
Approval of subcontractors employed by the subcontractor	172
Joint ventures	173
Subcontractor guarantees, warranties, sureties and insurances	173
Communication with subcontractors	174
Issuing drawings and instructions to subcontractors	174
Instructions	174
Subcontractor variations	174
Invoice procedures	175
Subcontractor's shop drawings	175
Subcontractor progress meetings	175
Combined subcontractor meetings – interface meetings	176
Payment of subcontractors	176
Back-charges	177
Clean-up work areas	178
Quality control	178
Samples, mock-ups and prototypes	179

Ensure the subcontractor makes money	179
Signs of trouble	181
Termination of the subcontract and the right to vary the subcontract works	181
Bribery and nepotism	182
Indigenous and local subcontractors	182
Protection of work and the work of others	184
Payment of subcontractor's employees	184
Maintenance and commissioning	184
Signing off subcontract work	185
Summary	185
Chapter 11 - Financial	**187**
Payment	187
Payment for unfixed materials	188
Payment retention	189
Non-payment	189
Day-works or cost-plus-fee	190
Variations	193
Costing variations	193
Change orders	194
Cost reporting	195
Cost to completion	197
Project budgets	197
Cost codes	197
Trading on claims	198
Contracts that are losing money	198
Basic site costing	201
Buy in from staff and feedback	201
Reconciliation of materials	201
Payment for goods and services	202
Vehicle log books and expense claims	203
Summary	203
Chapter 12 - Contractual	**204**
Contract documents	205
Checking contract documents	206
Laws governing the contract	207
The same contract – but is it?	207
Letters of intent	207
Documentation and records	208
Client's obligations	208
Designer's obligations	209
Managing contractor's obligations	209
Contractor's obligations	210
Types of claims	210
Informing the client	211
Drafting a claim	211
Acceleration	212
Force majeure	212

Concurrency of delays	212
Site instructions	213
Insurance claims	214
Resolving disputes	214
Notices	215
Liens	215
Terminating the contract	215
Liquidated damages	216
Summary	216
Chapter 13 - Completing and Closing Out the Project	**217**
Stages of completion	218
Completing punch lists	218
Quality control and other documentation	219
Certificate of occupancy	219
As-built drawings	219
Warranties and guarantees	220
Tying into existing services	220
Commissioning	221
Operation and maintenance, and training the client's staff	222
Spare parts	222
Plant and equipment	223
Financial	223
Closing out subcontractors	223
Demobilisation of staff, personnel and equipment	224
Clearing laydown areas	224
Termination of services and accommodation	224
Final account	225
Return of bonds and sureties	225
Archiving records	225
Lessons learned	225
Summary	226
Conclusion	**227**
Glossary	**228**

Introduction

Managing a construction project must be one of the most difficult jobs there is. To do it requires a large skill set, some of which can be learned by attending courses, colleges or universities, but others are only learned through experience, and often the hard way – learning from mistakes!

Skills required to manage a construction project

To manage a construction project requires technical skills and knowledge (for example an electrical project would require the Project Manager to have knowledge in the electrical field, while managing a road project would require knowledge of the mechanics of road materials, and a concrete project would require knowledge of concrete structures).

Apart from the technical knowledge, Project Managers may also need to know and understand the performance of various types of equipment, pumps, and machines. Some of these items are used in the construction process while others will be included in the facility.

The Project Manager should be able to read and understand drawings, see their interrelationship, and be able to visualise the construction process.

They should be able to plan the construction process to ensure the project is completed in the shortest possible time, and all the client's milestone dates and access requirements are achieved. The Project Manager must ensure all their staff and subcontractors understand the schedule, and deliver according to it.

The Project Manager has to procure materials and place subcontract orders, and then manage the delivery process and the subcontractors' performance.

Project Managers must ensure that none of the workers, visitors, or members of the public, are injured or harmed in any way while on the project, or by any activities related to it. They therefore, require a sound knowledge of safety standards, procedures and legislation, some of which can be complex and can vary between countries, clients and industries, and in some cases more than one set of safety regulations will apply. In addition, an understanding of the applicable environmental legislation is required.

They have to ensure that the construction project is delivered to the required quality standards, and in order to do this must be familiar with the project specifications, quality control documentation, testing procedures and tests required.

Of course, one of the most important aspects of managing a construction project is to be able to manage, work with, and interact with people. Construction projects involve a number of different parties, such as the client, managing contractor, subcontractors, members of the public, Design Engineers, Architects, managers within the company, project staff, workers, suppliers and their representatives, local businesses and building inspectors. Many of these people come from diverse countries, cultures, languages, socio economic backgrounds, and educational backgrounds. Project Managers may have to deal with an irate member of the public who is upset the contractor's delivery truck drove in front of them, discuss a design problem with the Designer, deal with the client's banker, or talk to a construction worker in a third world country who may never have had a formal job before (let alone, construction experience). All of these people have to be treated in an equal manner. No matter how frustrated you are, or how bad the day has been, there is simply no place on a construction site where you can hide from dealing with people and their problems. Project Managers have to use all their people skills to negotiate, persuade, and lead the people working on the project, while at all times remaining calm. They have to establish and maintain relations with the client, the design team and their own team. In all of this they also have to be a coach, a teacher and a mentor to their team.

Project Managers require a knowledge of industrial relations procedures so that labour harmony is achieved without jeopardising the project's productivity, or profitability. They need to be familiar with the company's procedures and requirements, specific project requirements, and possible union agreements, as well as legislation governing industrial relations which can vary between regions, states, countries and industries.

They need an understanding of basic financial principles, be able to read and produce cost reports, and ensure the project operates profitably.

An understanding of legal and contractual requirements, and obligations, is a necessity since they are required to put together claims and variations for changes that occur on the project.

In some cases, on large projects, there may be a team to assist the Project Manager, consisting of specialist resources, such as Planners, Quality Control Managers, Safety Managers, Engineers, Contract Administrators, clerical staff, and so on. But even in this fortunate position they must manage the processes, and ensure there is full compliance on the project, therefore they'll need to understand these processes. In some cases however, the Project Manager may be on their own, with only a Supervisor or two to assist, and they'll have to personally undertake all these functions to ensure the project runs efficiently. Sometimes the project is in another country or in a remote location, far from Head Office support. Your manager may be too busy to regularly visit you. In these cases the Project Manager will have to face, and deal with, all the challenges and problems on the project independently, relying on their knowledge and experience to make numerous, timely decisions.

A challenging and changing environment

On top of this managing a construction project is a twenty-four hour, seven-day-a-week operation. The project may not be working these hours, but I can't count the times I've received phone calls late at night, or on weekends, telling me that one of my workers has misbehaved, been injured or fallen ill, or that there's been a theft on the site, or a company vehicle has been in an accident or been stolen.

Furthermore, you have to contend with the environment. There could be prolonged dry periods when the site turns into dust, followed by periods of excessive rain when it may be impossible to work, parts of the work are damaged and flooded, or access roads become impassable. You could experience excessive windy conditions, high temperatures, or even extreme cold, while hailstorms and lightning storms could cause damage and disruption, all of which could prevent construction work from proceeding. Distant floods and rain may even impact the project by making routes impassable, or flooding the supplier's manufacturing facilities.

Then there are other external forces that can disrupt the project, such as national strikes which may affect the project directly, or perhaps disrupt suppliers. Delivery trucks may be involved in an accident, or the ship transporting vital equipment could sink, or alternatively be delayed by storms, incorrect customs paperwork or a strike at the harbour. There could even be industrial action on the project caused by another contractor which disrupts the work.

Client's vary tremendously, which means you may be working as a subcontractor to another contractor, or for a public or government organisation with strict rules governing the contract process, or a mining company, a major oil and gas supplier, or alternatively an independent inexperienced (project wise) home owner or small business owner. To complicate matters further, clients are not always consistent in their dealing with a contractor, something that's often influenced by the client's Project Manager and team. How you may have dealt with the client, and your relationship with them on a previous project, is rarely repeated, forcing you to learn different approaches and ways to work with them on every further project. Just as what was acceptable on one project may not necessarily be acceptable on the next one. Nothing should be taken for granted, even when the client is the same.

Furthermore, in the course of a project, clients and their Designer make changes, frequently revising drawings and adding scope. Clients may delay access to portions of the project, or their Designers may issue drawings late, or even drawings with insufficient or the wrong information.

Then just as you have mastered all the problems, finished your project safely, delivered it to a happy client and made a profit (hopefully), you have to move to a new project and do it all over again. Often the new project is for a new client in a different location, working with a new team, with different subcontractors and suppliers. The project could be completely different and result in you using different skill sets. No construction project is exactly the same as the previous one; each is unique, with its own set of problems and parameters.

Responsibilities of a Project Manager

All of these situations require the Project Manager taking action to prevent problems from occurring, and solving those that do occur. Decisions must be made every day, everything from minor decisions to major ones that could impact on the safety of the workers, or cost the project millions of dollars. Yet many of these decisions must be made in a hurry. The Project Manager has to manage risks on a daily basis, making judgement calls which can have large repercussions.

Project Managers also have to be salesmen selling themselves, and the company to clients and their teams, so as to enable the company to win and be awarded the project, and further work. They also have to positively influence the contractor's reputation and develop relationships with the client.

Furthermore they have to develop and maintain good relationships with the local community, and may have to work with and develop the indigenous people of the region.

Apart from running all of the above and ensuring the project is delivered on time, to the required quality, safely and profitability, it should be remembered the Project Manager also has a legal liability which extends across the whole project and the staff reporting to them. Failure to uphold this responsibility could result in you being prosecuted, and even going to prison should a person be injured on the project, or there be any environmental damage, and the project didn't comply with the legislation of the country or state, or the correct permits or licenses weren't in place. You may also be held liable for damage to third party property.

Indeed construction is not for the faint hearted!

Avoiding mistakes

And where does a Project Manager obtain the training to deal with all of the above? Well, the simple answer is, there is no formal training that can possibly prepare anyone to meet the challenges and stresses of managing a construction project. The best training is through experience, and hopefully a knowledgeable and experienced mentor willing to give their time freely to help and guide you through the process. The learning process will almost always result in mistakes. The important aspect though, is to learn from these mistakes, as well as the mistakes made by others around you. Equally important is to learn from your successes and the successes of others. This learning process will continue for your lifetime in construction.

I have been involved in the construction industry for twenty-eight years in both South Africa and Australia. The work has been mainly civil concrete, building and some roads work and minor mechanical works.

What I have found surprising is that no matter what the industry, project, location, or even the size of the project, the same mistakes are made time and again, and these mistakes are often a result of poor project management.

I've often seen schedules not followed on site, sometimes not even looked at by site management, or ones with major flaws submitted to the client. I have also been on sites where schedules have been updated incorrectly, or the client has been issued reports which are incorrect. As for resourcing of projects, or ordering of materials, this is seldom done correctly.

The contractual side of projects is often ignored, resulting in missed opportunities to be recompensed money, and contracts ending in contractual disputes.

Often the mistakes occur due to the fast nature of the projects. They may also be due to the scarcity of resources, which often means Project Managers are less experienced and the projects are under resourced, with insufficient required skills to support them. In addition, clients are becoming more demanding with their paperwork requirements, which take resources and time from an already stretched management team.

Some of the mistakes occur at tender stage. The estimate may be under-resourced, under-priced, used the incorrect subcontractors, or even have committed to an unachievable schedule. Sometimes this has nothing to do with the Project Manager, although on occasion they are involved and can influence the tender.

Of course the mistakes are not all directly related to the contractor's management. Sometimes the managing contractor's and the client's management teams could plan and execute the project much better than they do. After all, construction is a team effort and the most successful projects I've been on have been those where the contractor, the client and their managing contractor have worked together as one team.

How this book will help

In writing the book, I've tried to look at the time-line of doing the project and structured the chapters around this. Hence there's a chapter on planning the project, starting the project, schedule, running the project, safety, quality, human resources, equipment, materials, subcontractors, financial and contractual issues and closing the project.

Although much of the book refers to my experience in the civil and building side of the construction industry, many of these experiences will be typical for most forms of construction, whether it's a house, renovations, subcontractor work, electrical, plumbing all the way through to heavy mechanical and industrial construction. The problems are often the same, with a variance in the quantum of money when the project goes badly. I often feel sorry for smaller contractors on smaller projects because they may not have the support within their organisation to assist them, in fact some may just be one-man operations.

While I would not claim this was the bible of project management, I would hope that every reader finds something useful in the book. Many of the things included may seem obvious, but I can't tell you how often the obvious has been forgotten, causing a major problem on the project.

In various countries, within the different construction disciplines, and to some extent from one company to another, there are different terminologies used for the contractor's representative responsible for running a project. Sometimes this person is referred to as the Project Manager, Construction Manager, Site Manager or even Site Agent, but in general I have used the term Project Manager.

Also in this book I have assumed the project is a construction only project. There are various types of construction projects. The client may appoint Engineers and Architects to design the project and then contract the construction work separately, which is often done as one multidisciplinary package, or a variety of different packages relating to the different disciplines required, such as mechanical, electrical, civil, or building. Sometimes the client may make use of a managing contractor who will coordinate the project, including the design stage, as well as the construction packages and commissioning. In some cases, the client may include the design with the construction as one package, which becomes a design and construction project. On occasion the client may appoint the contractor to not only design and construct, but also to operate the completed project. Sometimes in addition the client may expect the contractor to finance the project. There are then numerous variations between these. The various arrangements all require some additional management and experience from the Project Manager.

This book will not give you the technical knowledge, nor will it give you the experience to become a construction Project Manager. There are courses, and even university degrees, run by specialists that will provide a Project Manager's qualification. It will, however, assist you to become a better and more proficient Project Manager, and help you to avoid some of the mistakes I've made, or witnessed, in my years in the construction business.

Chapter 1 - Planning the Project

The planning phase of a new construction project is the most important phase of the entire project. It's where Project Managers decide how they will construct the project, what resources will be required, and they prepare the contract schedule. Although this planning is usually done in a short space of time, the early decisions often survive the course of the project, and many of the problems encountered during construction are often a result of poor planning, and poor decisions, taken at the start of the project.

Case study:
One of the most successful projects I've been involved in was the civil construction works for a large minerals process facility. This involved the construction of concrete foundations for different structures and tanks which required over fifty thousand cubic metres of concrete, as well as the construction of several buildings.

Why was this project successful? Well, it was completed on time, constructed to a high quality standard, was within the client's budget, and was completed safely, with only one minor Lost Time Injury out of more than 1,600,000 man-hours we worked on the project. In other words, it was completed with minimal problems, disruptions and no industrial relations issues. Importantly, as a result we had a profitable project.

What was the reason for the success? Previously the client had always tendered their projects, and awarded their contracts almost entirely on the basis of the cheapest price. We had, however, successfully completed a similar, but smaller project, for the same client and client's managing contractor. On the basis of this success we were able to persuade the managing contractor and client to negotiate their next project with us, rather than requesting a number of contractors to tender for the works.

Negotiating the contract with us had a number of benefits for the client and ourselves:
- The client was certain they had a contractor who could deliver the project safely and on time.
- Since we had previous experience with similar projects we intimately understood the schedule and its critical paths.
- We understood what was required in terms of plant, equipment and formwork.
- We had a clear understanding of their requirements, so were able to plan around them.
- By being engaged on the project at an early stage we were able to:
    - work with their design team to ensure the design was constructible enabling significant savings on the construction duration
    - work with their planners to ensure the project schedule was realistic, and that follow-on contractors received the earliest possible access to the structures
    - influence some of the client's industrial relations policies

- - o develop relationships with key members of their team enabling the managing contractor, the client and us to work as one, anticipating problems, and solving them together at an early stage
- As soon as the financial and contractual negotiations started we knew we had an excellent chance of being awarded the project and we immediately started to plan the project, assembling our best team, ensuring many of the staff, including the Project Manager, had worked previously on similar projects.
- What's more, since the contract was negotiated, we were in a position to ensure all the required management resources were included in our price, so there were never any concerns that our price hadn't allowed for sufficient staff.

We and the client ended up in a win-win situation. The client took a major risk in changing their procurement process, which possibly increased the initial project costs, but overall ended up with a successful project, which in turn resulted in the final costs of the project being lower.

Unfortunately contractors don't always have the luxury of being involved so early. Clients often call for tenders on the open market, which can result in the client receiving bids from many contractors, most of which are working on and submitting several tenders at any given time. These bids may be unsuccessful, resulting in the contractor having insufficient work to keep all their staff busy, causing the contractor to release good, experienced people from their employ. Alternatively, a contractor may suddenly be awarded a number of contracts almost simultaneously, resulting in them experiencing a shortage of personnel and equipment. Add to this fact that the client may award the contract several months after the contractors submitted their tenders, which may mean when the project is finally awarded the suitable personnel the contractor had available when tendering have been placed on another project, or have left the contractor's employ. As a result, in a short period of time, the contractor has to employ new people, which often results in the contractor employing inexperienced or unsuitable personnel for the project.

In fact, the normal practice is that the client awards a contract and then expects the contractor to mobilise to site and start immediately (certainly within a period of two or three weeks). This leaves little time to plan the works. Contractors also have little influence on the client's schedule which may be overly optimistic and unachievable. In addition, the design of the major structures may already be complete leaving contractors unable to influence the designs which may not be construction-friendly.

In most construction projects the contractor's Project Manager isn't part of the tender phase or the post-tender negotiations. This is a pity, since if they were they would have played a part in planning the project and preparing the schedule, leading to a good understanding of the project, the project conditions and the contract document. Sadly, however the norm is that most construction companies only bring the Project Manager onto the project after it's awarded. This puts the Project Manager under a huge amount of strain because within a couple of weeks they must:
- figure out what the project is
- decide how to construct it

- know the requirements to construct it
- prepare a schedule
- submit the paperwork the client requires, such as manning histograms, organisational charts, safety plans, environmental plans, quality management plans, insurances and bonds
- appoint major suppliers and subcontractors
- source equipment
- find personnel
- prepare a project budget

The planning phase of the project is extremely critical, and the decisions that are made now will affect the success of the project and influence how efficiently the project runs. Poor decisions made during this stage could create difficulties later. Placing orders with the wrong supplier or subcontractor, ordering the wrong equipment, deciding to use cranes which are too small or too short, committing to unachievable completion dates, deciding on the incorrect schedule sequence, submitting the wrong paperwork or under resourcing the project will all have profound consequences on the project and may be difficult to remedy later. It's therefore critical the Project Manager works through all these items fast and efficiently. The extra hours spent on making good decisions at this stage will definitely be hours well spent.

A Project Manager needs to stay calm and focused, and work through the planning phase in a methodical and structured way to avoid becoming overwhelmed and making wrong decisions. Larger projects may require the Project Manager to assemble their staff so some tasks can be delegated, but they should always remain in charge.

The project could be civil, structural, or mechanical, it could be the construction of a house, office, new gas processing facility or simple home renovations. The project could be direct to the client or you could be a subcontractor. The project could be worth a couple of hundred thousand dollars, several million dollars, or even hundreds of millions of dollars. Despite this, the principles and problems of planning and executing the project are generally the same, with maybe only differing rules and complexities. Readers should therefore take what they need from what I've outlined below and adapt it to their needs.

I'll assume, at this stage, though that the project doesn't include the design of the structure.

Tender documents

It's essential the Project Manager obtains all the documentation the estimating department has accumulated on the project. This should include the full tender documentation, tender drawings, all correspondence during the tender period, the full tender submission, and post-tender correspondence, quotes received from suppliers and subcontractors, as well as all tender calculations.

I suggest the Project Manager makes a hard copy of all these documents so notes can be made on them and important paragraphs and facts can be highlighted. Some people may ask why in this day and age make a hard copy – why not save money and use the electronic copies? Trust me, the cost of the paper copies is minor in the scope of the whole project, and a hard copy is easier to mark-up and refer back to during the project. This documentation will become the basis of the project, so should be well read and understood.

File the documents correctly so it's clear what was received from the client and what was sent to the client. All correspondence should be sorted in date order with the most recent papers first. Minutes of tender meetings should also be filed. Tender drawings must all be clearly marked 'Tender'. These drawings could form an important part of any claim later should there be a change in scope.

Scope of works

The Project Manager and their staff must understand what they are going to build, where the project is located and what the client's requirements are. Only then can the Project Manager decide how to build the project and what resources, subcontractors and materials will be required.

So how does the Project Manager find out what they've been contracted to build? In the tender contract documents there will usually be a scope of works. This will give a broad outline of the project and what the contractor is expected to build. Hopefully the scope is clear. Some scopes of work are poorly defined by the client. If this is the case the Estimator should have clarified these ambiguities in the tender submission, as well as any assumptions that were made.

An important aspect of understanding the scope is to read through the tender submission, including its covering letter, which may specifically have excluded sections or items of work, or clarified aspects of the project.

Another place to obtain a general overview of the project is in the bill of quantities which the client may have issued with the tender documents. This bill should agree with the scope of works although this isn't always the case. As a rule of thumb, if the tender quantities include a structure, and the tender has priced that structure, it's probably safe to assume the structure is in the scope, even if that structure is not included in the scope of work. However, if the structure is not in the bill of quantities provided by the client, but is in the scope then the structure may not form part of the contract works. If there are any doubts, it's important the client is requested to provide clarification.

Tender drawings are also useful in providing an indication of the scope of works, although, they may be incomplete, for indicative purposes only, or even in some cases unrelated to the project.

This all sounds fairly basic? Well it is, but I have often been astounded at how many Project Managers do not know what's included in the project scope – even substantially through the project! You certainly would not want to be ordering expensive equipment which may be 'client supplied' or not ordering something because you believe it's not in your scope. I once had a Project Manager not order an item because he believed it was not in our scope to provide. The material then had to be especially air-freighted in at the last minute, at considerable cost, and resulted in a mile-stone date almost being missed.

Reading through the contract documents

As the Project Manager reads through the documents I suggest notes are made. One set of notes will be questions for the estimating team. Other notes will pertain to queries for the client. Make a note of the client deliverables and their due dates.

In addition, put together a list of thoughts on the construction process, and requirements needed to enable construction to get under way.

Important information required during the course of the project should also be noted, such as the:
- client's name
- contact person
- project insurance details
- requirements for the submission of the monthly valuations
- requirements for the submission of variations
- key dates
- liquidated damages or penalties

Consider preparing a standard form that can be used for every new project so the information is readily available and easy to refer to throughout the duration of the project.

Tender drawings

Going through the tender drawings can be quite a daunting and confusing task for the Project Manager especially since there may be several hundred of them. In addition, immediately after the contract is awarded the client will probably start issuing the construction drawings. It's essential the construction drawings are not mixed with the tender drawings. I suggest, at this stage, the tender drawings are referred to first.

The most important drawing is the 'Overall Layout'. Hopefully there is one. This should be filed and kept at the top of all the drawings for easy reference. I normally use a coloured pen to circle all the structures included in the project scope of work. If I cannot find some structures I make a note of these structures at the top of the layout drawing.

Next sort all the drawings into the different structures. At tender stage many standard detail drawings may be included for reference – some may not even apply to the project. File these detail drawings last. At this stage, when you're trying to absorb an overall view they can probably be put aside and studied later.

Carefully study all the tender drawings. Try to visualise and understand what the structures look like and what will be required to build them.

Hopefully between the scope of work, the drawings and the bill of quantities, (which may have been included in the client's documents or drawn up by the estimating team), the Project Manager will have a reasonable idea of what they'll be constructing. Obviously, if the Project Manager has worked on a similar project previously, they will already have an idea of what the structures look like and what will be required to construct them.

If you're in the fortunate position of having a full set of construction drawings (or certainly most of them), one set should be filed as a 'master set' and then also referred to during this planning phase. It's also essential to compare the construction drawings to the tender drawings and prepare a variation for any differences, such as an increase in the scope of work or changes to the specifications. These variations normally have to be submitted within a specified time period.

Access to site

It's important to understand the location of the project site. This may seem a logical step, but it's one thing to understand the general location and entirely another to understand the exact location in terms of access to site, site geography, and general local amenities. Are there road restrictions – low level bridges, poor roads, load restricted or narrow bridges? What are the access conditions to get onto site? Are permits required? Some of this information may be contained in the drawings, but more than likely it will be found in the contract documentation. Hopefully, during the tender process, the estimating team visited the site, and attended a site briefing where any issues and limitations regarding the site conditions were raised and explained.

Client provided services

Next the Project Manager must understand what the client is supplying and what is in the contract scope for the contractor to provide. Particularly important items are; accommodation, transport, water, power, toilets, offices, and materials. In addition, understand the locations of these services, and what their quantities will be.

Site conditions and constraints

The last important question is to understand the specific rules and conditions of the site. This may include; specific labour conditions, working hours, existing services, environmental restrictions, access to site, specific client requirements like numbers and types of staff (for example the client may specify in the contract documents the number of Safety Personnel required and their qualifications, or types of Quality Assurance staff, Planners, and Surveyors).

Hopefully, by this stage company management has allocated some staff for the project. If not, the Project Manager should at least have an idea of what key staff are required, and provided a requisition to company management or the human resources department to find and allocate them to the project. Key staff members are required to join the project team as soon as possible so they can assist with planning and preparations.

Tender handover

Having gone through the tender documentation and tender drawings, and with some idea of the project scope, the Project Manager should be ready for a formal tender hand-over meeting. This should happen for every project and could follow a set agenda.

Who attends the tender hand-over meeting will vary, and depend on the size of the company, and to a lesser extent, the size of the project. For small installations, or home renovations, the meeting may only include the Project Manager and the person who prepared the tender. In larger organisations it will be necessary to include the person who did the tender, as well as representatives from the different departments within the company which may include, but not be limited to: safety, planning, human resources, quality assurance, environmental, procurement, finance, and plant and equipment. Senior managers within the company may be included as required, as should key project team members.

The aim of the meeting is for the estimating department to explain:
- what the project is
- who it's for
- where it is
- start and completion dates
- basic thought processes in terms of how the contract was priced
- what equipment will be required and the source of this equipment
- numbers and types of staff required
- major suppliers and subcontractors
- contractual terms and conditions
- major risks and potential upsides
- logic behind any post-tender negotiations (savings offered, additional money added or acceptance of terms and conditions)
- specific project requirements

You may ask why I don't advocate having the tender hand-over meeting at the start of the planning process, before the process of understanding the project begins. A couple of reasons – firstly, it may take a couple of days to set a suitable time for the attendees, but more importantly because the Project Manager should go into this meeting with a basic understanding of what will be constructed and what the project is about. If a Project Manager has already gone through the tender documents and drawings they will be better prepared for the meeting and have questions to ask.

Site visit

At this stage it would be useful if the Project Manager could visit the site, especially if they weren't part of the original tender visit. This will give an overview of the project and general site conditions. Items that will become clearer include: the ground conditions, general terrain, other contractors working on site, location of services, the position of the lay-down area, site specific requirements, and the site access.

During the site visit, the Project Manager should take the opportunity to meet the client's staff on the project. By talking to them a better understanding of the site conditions and constraints can be obtained along with learning what the client's team consider to be important aspects of the project, such as their key deliverables, important dates, and risks they perceive may affect the project.

Often because other contractors are now on the site, access conditions may have changed from those previously envisaged. After visiting the site the Project Manager may, possibly end up changing ideas on the method of carrying out the work.

Planning the execution of the work

By now the Project Manager, together with their team, should have sufficient information to plan the construction of the project. They have heard the thought processes from the estimators, and how they envisaged the structures would be constructed. They have seen what the client has proposed, and how the structures have been designed. The construction team may choose to use these same methods, or they may develop alternative methods, using the information they now have, which wasn't perhaps available to the estimators.

Case study:
One of my projects involved the repair to a rail bridge in Mozambique that had been severely damaged during floods, causing the last four bridge spans on either side of the bridge to be washed away. As a temporary repair, to get the railway line operational, the missing bridge spans were replaced by filling the gap with soil, and placing the rail lines on this filled material. This, however, restricted the river flow, and was at risk of being washed away in the next flood.

While replacing the eight bridge spans the railway had to be kept operational. The client's engineer proposed a method of constructing the bridge spans from concrete in pre-cast sections which would then be craned into position. A major problem for us was that a large crane was required (in excess of three hundred ton capacity), and since the project site was in a remote area this was going to be costly. Also, while the earth was being removed and the pre-cast sections placed, the railway would be shut for many hours, or even days.

We developed an alternative solution which did not involve removing the railway track until literally the last day. We installed steel-cased, concrete-filled piles through the earth embankments. Later when we excavated the temporary earth-fill embankment, these piles were exposed and became the columns for the new bridge. While the railway was in operation, we constructed the bridge spans next to, and parallel to the rail tracks, on concrete beams, which connected to the new bridge piers formed on top of the new piles. After completing the new spans we shut the railway down for thirty-six hours, removed the railway track and then slid the concrete bridge spans onto the new bridge piers using small jacks to pull the spans into position. After locating the spans in their correct position, we replaced the railway track, and reopened the railway. We then removed the earth, used as the temporary fix after the flood, from under the new bridge spans, restoring the river bed to its original pre-flood shape.

This construction method not only minimised our costs and saved time, but it also minimised the time that the client could not utilise the railway.

Remember you don't generally have to use the construction methods or equipment that the client or your estimating team proposed. They are only guides.

Schedule
A project schedule must be developed by the contractor and approved by the client. Many Project Managers seem to believe that the schedule is not their responsibility. Wrong! I would ask that all Project Managers read Chapter 3 of this book – it's probably the second most important chapter after this one. The schedule is a useful tool for the project.

The schedule:
- is important from a contractual point of view since it will dictate when penalties or damages may be imposed by the client on the contract
- will assist with any potential delay claims and costs
- governs the order of construction

- dictates the resources required for the project (the type and quantity and when they are required)
- indicates when materials should be ordered
- shows when information is required from the client
- is used to measure and record progress

The Project Manager may be in the fortunate position of having access to a Planner to prepare the schedule. The Planner may even be allocated to the project full-time. However, it must be noted that many Planners are just data in-putters and do not come from a strong construction background. They may understand the intricacies of the schedule system, but often don't have suitable construction experience. Consequently the Project Manager must stay in control of the schedule dictating how the contract will be constructed, in the correct order and utilising resources efficiently, since they are responsible for the success of the project, and the schedule is integral to its success.

It's important to remember the schedule prepared and submitted at this stage will become the contract schedule. For this reason the Project Manager should devote a considerable portion of time to its preparation, ensuring it's both workable and achievable. A poor schedule may result in:
- the contractor being unable to submit extension of time claims due to delays caused by the client
- the contractor having inadequate and incorrect resources for the project
- the contractor committing to completion dates which cannot be met
- the client imposing penalties for late project completion

It's normally a contractual obligation to produce the contract schedule within a specific time after award of the contract. It's also in the contractor's interest to produce it as soon as possible since the schedule will dictate the access date to site, as well as the 'information required' dates. It's amazing how often the clients don't meet these deliverables, yet if there's no approved contract schedule it's more difficult to claim against the client for any resulting delays.

The tender schedule should be used as the basis for the contract schedule. The client will normally insist that the durations in the tender schedule, and the dates committed to during the tender stage, remain unchanged. However, now is the last opportunity to set a schedule with realistic dates, and the Project Manager must endeavour to do so.

The Project Manager usually requires construction drawings to create the schedule. I often hear Project Managers claim they are unable to create a schedule because construction drawings have not been issued. If these haven't been issued, or are insufficient, then the tender drawings will have to be used, and the schedule submission must reference which drawings were used.

It should also be noted that the drawings issued for construction may be different from those issued at tender stage. The tender drawings may only have been schematic with simple lines, which might not be distinguished as being concrete or steel. The construction drawings could show additional details which may not have been envisaged at tender stage. These details can often dictate completely different methods of construction and durations for the activities, yet I've seen Project Managers blindly duplicate the tender schedule without considering how the construction drawings differ from the tender drawings.

Case study:

Some years ago we constructed two large concrete silos, sixty metres high and eighteen metres in diameter. At tender stage the drawings weren't very clear and showed one major concrete slab in the silo about fifteen metres above ground level. Below this slab were a number of vague lines that possibly showed other floors and levels.

Having constructed many silos previously, I knew that below the main concrete slab were generally a number of structural steel floors holding equipment for the silos. These steel floors were normally part of the mechanical installation by a specialist mechanical contractor employed directly by the client and were not part of our scope. In our tender schedule, included with our tender, we allowed for only one concrete floor slab at the fifteen metre level, believing that any other floors weren't part of our scope.

When we were issued with the construction drawings for the silo, the lines on the tender drawings now became two additional concrete floors. To me this was a major addition to our contract and their construction added an additional six weeks to our schedule. Yet the Project Manager responsible for the project submitted the contract schedule to the client, adding in the additional concrete floors without changing the overall schedule duration. Unnoticed this could have significantly increased costs, as not only would we be on site for the additional six weeks, but would have been unable to meet the contractual completion date committed to in the contract schedule.

Fortunately we noticed this oversight and were able to revise the schedule to include the additional time required. We then submitted an extension of time variation for six weeks, arguing that the two concrete floors were additional scope. Initially the client's representative was unhappy, but thankfully we could demonstrate that at tender stage the additional floors were not clearly shown on the drawings, so our tender schedule had not allowed for them. We could also demonstrate it was the norm for these floors to be fabricated in steel, and installed as part of the mechanical works contract once all the concrete works were completed by the civil contractor. Consequently we were able to get our extension of time of six weeks with additional costs.

This example highlights the importance of submitting a schedule with the tender submission, even if this hasn't been requested. It not only demonstrates that the contractor has thought through the project and has an understanding of how it will be constructed, but it can also protect the contractor during the contract, as demonstrated above. Always review the contract schedule before it's submitted, using the current information.

An important aspect of the schedule is the resourcing of the activities (personnel and equipment). The Project Manager needs to know:

- the number of workers
- when they are required
- their duration on the project
- their skills

If the tender schedule has been resourced correctly it can be relatively easy to resource the contract schedule. However the tender schedule shouldn't be blindly followed without reviewing it.

I would say that two of the biggest problems I've encountered on projects are either the project has had insufficient resources, or the resources have not had the required skills.

Case study:

Some years ago the company I was working for bought out another construction company. One of the projects they were working on was a concrete structure which consisted of walls twenty-six metres high and involved large amounts of formwork, reinforcing and scaffolding.

The contract was a classic example of what happens when you have enough resources, but the wrong mix. Most of the workers were highly qualified form-workers, however with insufficient assistants to help them. This resulted in relatively expensive resources doing simple tasks. The qualified form-workers felt they shouldn't be carrying out menial tasks, so these were performed reluctantly and inefficiently. There were also insufficient reinforcing-hands which meant that the wall formwork was delayed because the formwork couldn't be closed up until the reinforcing was completed. Furthermore there were insufficient scaffolders to build the access-scaffold for the reinforcing-hands and the form-workers, so again, more delays. To compound the problem, for some reason the project had an excess of leading-hands who weren't interested in doing actual work, only wanting to supervise.

So how had the project got to this state? Well, firstly, before starting nobody had planned the project and worked out which resources were required. They had simply asked for a number of people and they got that number. Even after the project had been running for several weeks, nobody stopped to analyse where the pinch points holding up progress were. The problems were not so easy to resolve. It was easy to get more trade assistants, reinforcing-hands and scaffold erectors, but what do you do with the excess leading hands and form-workers?

Even if you're doing a simple project and have only done a hand-drawn schedule, it would be a useful exercise to create a simple resource histogram showing the different trades and the number of personnel required for each trade.

Subcontract or self-perform

An important decision required in the planning stage is whether the project will be constructed using workers employed within the company (often referred to as 'self-performing the works'), or if portions, or all the work, will be performed by using subcontractors. This will be influenced by a number of factors:

- What the company policy is regarding subcontractors. Some companies will subcontract out all the works, while other companies will do as much of the project as possible and the Project Manager will have to abide by this policy,
- Some parts of the project may be specialist in nature and your company may not have the specialist resources within it to carry out the works making it an easy decision to subcontract. However, in certain cases it may

be difficult to find a subcontractor to do these specialist works due to the location, size or project cost, or their workload. If a suitable subcontractor cannot be found the contractor will have to give consideration to doing the specialist works with-in the company, forcing the Project Manager to employ suitably qualified personnel to carry them out.
- In some cases a company which normally self-performs the works may be too busy on other projects, resulting in insufficient resources being available for the project. In this case, by subcontracting portions of the work, there will be a reduction in the quantity of company resources required for the project.
- If the project is in a remote region, or another country it may be more cost effective to employ a local contractor who may have accommodation, facilities and resources in the vicinity of the project. In addition, there's no need to transport employees long distances to their home base, or pay additional allowances as an incentive to persuade personnel to work in that remote area far from their home base.
- In some cases, there may be a project requirement to engage local or indigenous contractors on the project, in which case portions of the project will have to be subcontracted out.
- Subcontracting portions of the work passes some of the project's risks to these subcontractors, lessoning the contractor's risk exposure.

There are different subcontracting methods. If you choose to subcontract out a section of works in its entirety, the subcontractor will then supply all the supervision, workers, equipment and material to carry out the works. Alternatively, the subcontractor can be contracted on a labour-only basis, where they only supply labour and the main contractor supplies the supervision, materials and equipment. There are then variations between these two forms of subcontracting.

How and what portions of work is subcontracted out can influence the project schedule, the numbers and type of staff required, and the numbers and type of workers and equipment needed.

Alternative construction methods

At this stage it's important to consider alternative methods of construction which may reduce costs, shorten the schedule, reduce resources or improve the quality. As an example, on a concrete or building project pre-cast elements could be used. These can save time on the schedule and eliminate some of the manning required. Pre-cast elements can be constructed on the project-site or fabricated off-site, with the fabrication done by the contractor or subcontracted out.

To reduce the amount of work performed at heights, or the number of skilled personnel, building projects make use of modular construction to fabricate sections, or rooms, off-site. Structural steel or piping contractors sometimes fabricate modules, on or off-site. These modules are then placed in position and connected together.

These methods of construction can save significant amounts of time, reduce the safety risk and the number of workers required on the project. Disadvantages may be the high costs to transport the modules to the site, and the required heavy lifting equipment to handle them, which can be costly and not readily available. Projects with many repetitive sections will benefit more from utilising pre-cast or modularised components than others might.

Consideration can also be given to the use of alternate materials which may be more readily available in the project area, or those that offer significant cost and time advantages if, for example, they are lighter or come in longer sections, which saves on their installation time.

Client approval will be required to use alternative construction techniques or materials. Some may also require the client to make changes to their design or for you to employ your own design engineer to do so. The costs required to modify the design must be taken into account when considering the benefits of the changes. In addition changes to the design or the materials may result in the contractor taking on some of the design risk for the project. Again, the cost of accepting this risk must be carefully considered. Sometimes these additional costs result in the alternative no longer being viable.

Alternative designs

It's normally difficult to influence the client's design at this stage. However, it's something that should be considered.

The changes may be simple and readily made, for example changing concrete column sizes so the columns conform to the formwork the company owns or has available. Reinforcing layouts or structural steel detailing could also be altered to make construction easier. More radical changes could be considered like changing concrete structures to steel structures.

Even minor changes can have significant time or cost benefits to the project, although many clients are reluctant to make changes due to additional design costs, however, by demonstrating the benefits to the project, such as improved safety or time, the client may reconsider and make the changes. In some instances the additional costs of employing your own design engineers to make changes may be insignificant compared with the savings the improvements make to the constructability of the project.

Formwork

There are many different types of formwork. The decision on what type to use could be influenced by:
- what the company has available
- what type is best suited to the structures to be constructed
- what is available in the vicinity of the project
- the price of the formwork
- what the Supervisors and workers are familiar with
- the skills of the work force
- the number of times it can be used on the project
- the finish the client requires

Consideration must be given to whether the project will purchase or hire the formwork. If the formwork will be reused many times maybe it's better to purchase it, and specialised formwork will probably have to be bought for the project.

I've had many projects make the wrong decision regarding the type of formwork, resulting in formwork being purchased that was slow, cumbersome, heavy, and requiring more labour to assemble than other systems. If more thought had been given during the planning stages, weighing up the pros and cons of the different systems, costs would have been reduced, as well as frustration.

Vertical access

Many construction projects require vertical access which may vary from just a couple of metres to very high structures. This access can be provided by building an access scaffold which may be fixed or on wheels, by using mobile access platforms such as scissor lifts, by hanging or suspending scaffolding from the sides of completed structures, and even in specialised cases using helicopters.

The most important consideration when deciding on a system is safety, and the system selected must be safe and meet the local legislative requirements as well as any special requirements the client may have. Other considerations are:
- what systems are available
- the type of access required and its height
- the costs of the different systems
- how long the access is required for
- the site layout and topography
- what systems Supervisors and workers are trained to use
- what impact the access may have on other activities and the schedule

Manning

Once the scope, schedule, construction method, materials and the project constraints are known and understood, the quantity, skills and types of workers can be determined, and the project can focus on recruiting personnel.

Often the recruitment of personnel, and in particular craft and general labour, is left entirely to the human resources department, the Site Administrator or a Supervisor. The Project Manager should ensure whoever is recruiting understands the type of skills and experience required. Ideally the curriculum vitae of the blue-collar workers should be reviewed by the Foremen or Supervisors working on the project since they will be the ones working closely with the workers. They are also best equipped to understand the skills and experience.

The staff recruiting for the project must also be familiar with the conditions of employment, especially since the conditions for your project are probably different from other contracts the company is currently working on. These conditions should be clearly explained to potential recruits because there's nothing worse than people arriving on a project with certain expectations when the reality is actually quite different. New employees need to understand rates of pay, employment conditions, site hours, site rules and safety requirements applicable to the project. In addition, the personnel must be familiar with the conditions relating to how and when they will go home on break, and what length the break will be.

Staffing

Only after the numbers of craft persons and labourers has been determined will it be possible to finalise details of the required supervisory, management and support staff.

The numbers of staff and their positions depend on a number of factors, such as the:
- number of work areas
- numbers of workers
- available skills of the workers
- complexity of the project
- relative locations of the work areas and their distance apart
- client requirements
- contract reporting requirements
- skills and experience of the available staff

Staff numbers also depend on the different skills and disciplines required. For instance in the civil industry Supervisors may only be skilled in earthworks, while others are skilled in concrete works, electrically orientated, or more suited to building trades. If you can employ a Supervisor skilled in a multitude of disciplines, there's often a reduction in the number of staff required.

To assist with calculating staff requirements prepare a project organisational chart. This shows the different levels and types of management, the reporting structures, as well as the different roles and responsibilities. In addition the chart could indicate whether the staff are required full-time or shared between projects, and if they are Head Office or site-based positions.

I've often been guilty of under-staffing my projects. The risk with that is:
- if a member of staff falls ill or resigns there's nobody who can take on an additional roll
- tasks are rushed resulting in mistakes and items being overlooked
- staff work under pressure, and long hours, which may lead to them resigning

Obviously having too many staff leads to:
- higher costs which were not allowed for
- staff becoming unhappy due to boredom and feeling underutilised

It's necessary therefore, that careful thought goes into correct staffing to achieve the best balance of efficiencies and costs. In Australia for instance, the cost of a Supervisor is not much greater than some blue-collar workers, so it may pay in some cases to have more supervision just to improve the efficiencies of the workers.

Remember also that if the project has to work nights, or weekends, additional staff will be required to supervise and manage these shifts.

Once the organisational structure has been decided, it should be compared to the organisational structure prepared at tender stage. Hopefully the structures will be similar. If, however, they are substantially different the Project Manager should analyse where and why there are differences.
- Have you misunderstood the project?
- Are you being too conservative and allowing more staff than required?
- Has the scope changed from tender stage? In which case a variation could be submitted for the additional staff.
- Maybe the tender was misunderstood and too few staff was allowed for at tender stage?

- Maybe the overall schedule has been shortened, in which case, more staff may be required for a shorter duration.
- Maybe the client has awarded the contract late, revising the schedule to become compressed? If so, a variation could be lodged because the client has caused the project to be accelerated, resulting in the need for additional staff.

If the number of staff required is more than the staff numbers estimated at the tender stage the Project Manager will have to re-budget the project to allow for these additional costs.

Also after completing the organisational chart it's a good idea to fit the staff required into a time-line, since not all the staff will be required from the start to the end of the project. There's normally a build-up of the staff at the beginning, which depends on the histogram for the blue-collar workers, and when each section of works, or trade is scheduled to start.

Once it's determined what staff are required, when and what skills, attributes and experience they should have, then the Project Manager can begin to source them, either from within the company or from external sources.

The best person for the job can depend on many things. Principally of course it's about knowledge and experience. Preferably you would have worked with some of the staff before. Certainly the most successful projects I've run were ones that utilised staff I'd worked with previously. I knew what they were capable of doing and understood their strengths and weaknesses. I was, therefore, able to place them in the positions that were best suited to them and their capabilities, and in knowing their weaknesses I was able to ensure there were sufficient support structures in place around them to compensate for this.

While we all like to work with the strongest members of staff this isn't always possible, and for some projects it's inappropriate to use them.

Case study:

One of my projects involved a large amount of demolition work. We couldn't start demolishing the follow on section until we had reinstated and rebuilt the previous one. This was a slow process. We had to work in many different areas across the whole facility at the same time and each area required a separate Supervisor. The work was fairly simple, but there were big concerns around safety because it was in a live factory that was noisy and messy and used a variety of dangerous chemicals. If I had put senior concrete Supervisors on each section of works they would have been frustrated with the slow pace of the works, and since there were no large volumes of concrete there would have been little job satisfaction for them. Senior Supervisors would probably also have come with large teams of workers with inappropriate skills for the tasks.

I, therefore, only allocated one senior Supervisor to the project who I knew was patient and wouldn't be frustrated or unhappy with doing the small bits of work. I also knew that he would be willing to assist and train junior inexperienced Supervisors, so I then utilised fairly junior Supervisors for the balance of the areas of work. Most of these Supervisors were newly promoted to the role, happy to work on their own little section, and only had to control a small team in a limited defined area. They also had a senior Supervisor on the project who was willing to give them help and advice when required.

The project was a success because we had a happy team, despite the project not being the most rewarding in terms of constructing interesting and new structures. The new Supervisors were learning and growing on the project and all of them developed into fully-fledged Supervisors, later moving to bigger roles with larger teams on other projects. The senior Supervisor had job satisfaction because he was able to coach and mentor the junior Supervisors. The project was a financial success due to junior Supervisors earning considerably less than a senior Supervisor. Importantly the client was pleased because we had a Supervisor on every section of the works and a good safety record.

Suitability of staff is also not just about the experience of a person. An important aspect on a large project, where there may be a number of Supervisors and Engineers, is that each person should be a team player, willing to share plant and equipment, or to help out in other areas. I have often had Supervisors who thought they should have plant and equipment allocated full-time to their sections of work and that they should take priority when materials arrived on site. Some of these Supervisors even had their own store which they controlled as if they had personally bought the materials. This practice is costly and causes dissent with other staff.

On occasion a Supervisor possibly will, due to the size and complexity of the project, have to report to a more senior Supervisor. This might not suit some who have been used to working alone on smaller projects. Again, this can lead to problems.

Some people may well just be good at doing certain tasks or kinds of work, and there's no denying that it's helpful to the project if you can use a person who has done a similar task before. The risk, though, is that a particular person keeps being given the same role, for example, using the same Engineer to look after quality, or the same Supervisor to supervise the concrete mixing plant. This could suit particular individuals, but in general people become unhappy and disillusioned if they are continually given the same or similar tasks. It's also not good for their future careers since they aren't learning anything new.

Many times you don't know the person or their capabilities because they have not previously worked for you. In this case you may have to ask the person they reported to on a previous project for feedback. Reference can also be made to the person's curriculum vitae which should give some idea of their experience and qualifications.

Inductions and site access requirements

The tender document should detail the specific site access requirements for personnel to work on the project. These may include, but not be limited to any of the following:

- specific medicals and particular service providers for these medicals
- specific training courses and definite training providers
- verification of competency certificates could perhaps be required for plant and machine operators which might have to be done by a client-specified provider, and some clients even specify that the ticket is valid for a limited

time and for the exact model of machine the operator will operate
- project specific inductions
- specific approval processes workers must pass before being allowed to access the site

It's important this process is understood by the Project Manager, recruitment staff and workers. Often completing the process to access a project can be lengthy and I've known projects where it's taken more than four weeks to get a person onto the project. Failure to understand the process and the submission of incorrect paperwork can extend this process even further. I've had projects where workers have been sent on the incorrect training or been trained through an unapproved provider. This not only becomes costly and delays getting the personnel to site, but also creates frustration with the personnel involved.

Choosing the plant and equipment

During the planning and scheduling stage of the project, thought should be given to the type and quantity of plant and equipment required. The Estimator would have had some thoughts on this matter when preparing the tender. You may, though, decide to use different plant and equipment if the items the estimator planned to use are no longer available, if other equipment has subsequently become available, or if the methodology of construction has changed from the tender stage.

I would suggest that the contract schedule is resourced with plant and equipment since this will provide the project team with the information of the type and quantity of equipment required, as well as when and for how long.

Check the availability of the plant and equipment both from within the company and from external hire companies. If the chosen equipment isn't available alternative items may have to be used, or if there isn't a suitable item available (for instance a large heavy lift crane), you may have to rethink the project's construction method, which may influence the schedule and staffing requirements. It's, therefore, vital the availability of people and equipment are considered when the schedule is being prepared.

As soon as the choice of equipment has been made an order should be placed to secure it.

Plant and equipment is not just about sourcing the major items of equipment required to construct the project, but also includes minor tools. In addition the Project Manager must ensure sufficient equipment will be available on the site to enable the construction of the site facilities. On occasion I've witnessed site facilities arrive at the start of the project and the site personnel not have equipment to offload and erect them.

Site establishment

Once the peak staffing for the project has been determined the offices, stores and facilities required for the project can be planned. Depending on the contract the client could be providing some of these, or possibly even require additional offices or facilities for them, their managing contractor or other contractors on the project. Also always check quotes from subcontractors to see if there's a requirement for you to provide facilities for their staff.

If facilities are to be provided for the client or the managing contractor these should be in accordance with the special requirements outlined in the contract document. Often the client may require offices that are a different size or specification to what would normally be provided, or different to the contractor's standard offices.

There may be restrictions on the type of temporary facilities the contractor can construct, as well as their locations and size. The facilities should be located close to the works, but not in the path of the construction work. Many building sites are confined and unless the facilities can be sited on a vacant lot next to the project site it's often difficult to locate them in a position that will not, during the course of the project, be in the way of some structure, road or landscaping which will mean the facilities having to be moved during the project. Ensuring there is an alternative plan for when the facilities have to be moved, and that they are easily relocatable, will lessen the disruption to staff.

If you're lucky there may already be facilities nearby, or old buildings on the site, which could be used. This is beneficial at the start of the project since the construction team can begin working immediately from the offices, without waiting for temporary offices to be built.

Site office requirements

Temporary facilities must comply with the local building codes, and be designed and constructed to withstand the local weather conditions. For projects in tropical cyclone regions, even when construction is outside of the cyclone season, the facilities are required to be designed for cyclonic conditions. In the past, I've been involved in a project constructed outside the storm season but, due to additional works and delays, the project was completed well into the cyclone season.

I recommend that all facilities are lockable. In addition, lockable storage should be provided within them for sensitive documents such as employee records, tender documentation, and sensitive costing and financial information which should be kept from the client, general employees, subcontractors and suppliers.

Offices should be planned to provide sufficient space for staff to carry out their work. The amount of space allocated to each staff member may vary according to the job they are doing. A Project Manager should generally have their own office to deal with confidential meetings with the client, staff members, company management and workers, and it should be large enough to accommodate a couple of chairs for visitors, to open up the project plans and to store the required project documentation. It's essential that many of these meetings can be held without being overheard by the other staff. Safety and administration managers may also require facilities to hold confidential discussions with employees. It's, however, not necessary for every member of staff to have their own office, although each staff member should have an allocated desk where they can work and store their project documentation. There is nothing worse than staff having to share facilities since it impairs the productivity of all.

I always think a conference room is a necessity on most projects. The size of the room will depend on the size of the project, it should be large enough to hold meetings with the client (if this is a contract requirement), subcontractors and also staff inductions and training sessions.

Consideration should always be given to the layout of the offices. For instance, the Site Administrators have enquiries from employees on a regular basis, therefore, they should be located close to an external door, so employees don't walk through the entire office. Contract Administrators may have sensitive information so their offices should be the least accessible, and the Supervisors and Foremen should have offices as close to the work area as possible.

The offices must have sufficient light and ventilation. In areas of extreme heat or cold the offices may require heating or cooling systems, and insulation in the walls and roof. Sufficient tables, desks, chairs, filing cabinets and cupboards should be provided to accommodate all staff and their documentation. Although you don't want to encourage staff to be in the office when they should primarily be in the work areas, staff should have a reasonable level of comfort while working in the office to achieve maximum productivity.

Worker facilities

Normally it's the contractor's responsibility to provide the worker's facilities like, toilets and eating areas, unless the contract document calls for the client to do so. The type and size of the facilities depends on the:
- number of the contractor's employees
- number of the subcontractor's employees
- specific contract conditions with subcontractors
- client requirements and project contract conditions
- worker and union agreements
- state or country legislation

Sufficient facilities should be provided to cater for the peak manning of the project.

Eating areas should have sheltered seating and tables, with an adequate supply of drinking water and washing-up areas.

Provision of utilities and site services

To enable construction work to proceed contractors require utilities and site services at the project site. The amount and type of service depends on the type and size of construction works. These services normally include power, water, sewer, telephone and data and, possibly, compressed air.

On many projects, the client is responsible for supplying the service connection to a location specified in the contract document, and it's up to the Project Manager to ensure the client meets these obligations. I've had experiences where the client has not provided the services on the contractual date, or provided them to a point further than specified in the tender, or in an insufficient quantity.

Even when the client is responsible to supply the services, the contractor normally has to provide the electrical transformers, meters, cables, valves and pipes to get the supply to their project offices, and to areas where they are required. In addition, consideration may have to be given to providing water storage tanks and back-up power sources should the supply be interrupted. All the material and the subcontractors required to carry out the installation, should be organised during the project planning phase.

Often the contractor has to source and provide the services necessary for the construction work. For projects located in cities and towns the provision of services can be relatively straight forward and simply require that the relevant applications be lodged with the local authorities and the payment of a deposit and connection fee. Sometimes, though, the authorities can take several weeks, or even months, to provide the connection. To avoid delays, the application should be lodged as soon as possible, still it may be necessary to provide temporary arrangements to supply water or power until a connection is provided.

It's essential to confirm that the connections installed will provide the quantity that the construction works requires. To ensure this the peak consumption will have to be calculated. These calculations need to take into account the requirements for commissioning, as well as the subcontractors' requirements. It's also important to understand the quality and reliability of the service.

Some projects in remote areas require the contractor to make elaborate plans to obtain water. These may include pumping water from underground aquifers, rivers, dams or even the ocean. It should be noted that many of these options require a permit from the relevant authority. Sometimes desalination or water purification plants are also required to treat the water and make it potable. There are package units available for these processes.

In many cases there will be various solutions to providing the utilities. It's important the different options are clearly analysed to ensure that over the lifetime of the project construction the cheapest and most reliable solution is chosen, giving due consideration to the duration of the works. Remember that rivers, dams and underground aquifers can dry out, so careful thought should be given to the quantities of water required, as well as the time-of-year in which the work will be undertaken.

Most areas have mobile telephone coverage which includes mobile data coverage. Check the strength of this coverage and compatibility with the systems your company uses. I've known a managing contractor arrive at a project, yet have no communications for several weeks due to incompatible systems. Where mobile coverage is weak consider putting up aerials, installing boosters, or in remote locations using satellite communications. This can be fairly expensive, but poor or a lack of communications can severely hamper productivity and the additional cost of installing a good communication system of sufficient capacity will more than pay for itself over the course of the project.

The use of portable radios will reduce the cost of telephone calls, as well as improve productivity and safety. Ensure the portable radios are robust, reliable, having good battery life and are set to a radio frequency that you are licensed to use, and is not being used by others in the vicinity. To improve the range of the radios on large projects it may be necessary to install repeating stations. On a project with many radios in use consider utilising more than one radio frequency - especially where radios are used to control crane lifts or other high-risk tasks.

Accommodation of workers and staff

An important aspect on some projects is finding accommodation for the project personnel. Obviously for some projects, located in the larger towns, there are enough locally-based workers thereby negating the need to import and accommodate project personnel, although this may depend on a number of other factors, like labour and union agreements. On some projects the client will supply accommodation for staff and workers, while for others only for the workers. If the client is supplying the accommodation it makes things simpler and the only concern is to understand the procedures the client has in place for booking in employees.

When the contractor has to provide accommodation for their workforce, finding suitable accommodation will depend on location and availability. It may involve renting rooms in existing camps, renting houses in nearby towns, or for remote locations temporary accommodation camps may have to be constructed.

The accommodation provided must:
- be located as close to the project as possible to limit long commute times
- meet the relevant legislative requirements
- comply with the current labour agreements and union agreements
- have sufficient toilets, showers, cooking, eating and recreational facilities
- be structurally safe, providing protection from the elements
- be secure
- accommodate the peak number of personnel expected on the project (depending on the subcontractor contracts this number may include some of the subcontractors' personnel)
- have water, power and sewer utilities designed and installed to meet the expected numbers, and I would allow for an extra 10 - 20% additional capacity and for extra use of power during periods of extreme cold or heat

If workers have reason to be unhappy with the provided accommodation it can lead to poor morale, which will impact the productivity on the project and, in the worst case scenario lead to industrial action. It's easy to underestimate the amount of accommodation required for a project and half-way through the project find yourself scurrying around trying to install additional rooms, putting a strain on the facilities and utilities provided, causing further problems.

Depending on the labour and project agreements meals may also have to be provided to the personnel. This is a minefield that I don't want to enter, all I can say is that great thought and care needs to be put into this exercise.

Ordering materials

During the project's planning phase it's necessary to begin placing orders for materials, including those required for the construction of the temporary facilities, various consumable and stock items needed during the course of construction, items that will be used in the first stages of construction as well as items with long-lead times. It's wise to order sufficient quantities of consumables to ensure there will always be a supply available on site.

Safety and environmental preparations

Safety is covered in detail in Chapter 5. However, it's important that safety forms part of the planning phase of the project, since it can affect the construction process, sequence, schedule, and the plant and equipment that will be used.

Before work can begin on the project, safety and environmental management plans must be prepared. This is a process that many Project Managers delegate to their safety staff or the company safety department, unfortunately this leads to many plans being prepared by people who might not be familiar with the project and the particular hazards associated with it. For this reason it's important that the project team involved with the various construction activities assist with preparing the safety and environmental plans so as to ensure they adequately address all of the risks. These plans also need to take cognisance of the client's overall project safety and environmental plans.

During the planning phase, risk and hazard assessments will need to be prepared for the tasks happening during the first few weeks of accessing the project site. These assessments apply particularly to tasks relating to delivery of materials, construction of the site facilities and the installation of project services. On occasion I have experienced the first delivery trucks arriving on site but being unable to offload because there was no risk assessment available for the task.

In addition, it's essential to procure the relevant safety registers and stationery, ensuring they are available on the site at the start of the project.

Safety equipment must also be procured and available from the start of the project. This would include, amongst other things; first-aid boxes, fire extinguishers, inspection tags, safety signage, safety posters, and barricading. There must also be, on site, sufficient stock of personal protective equipment, such as boots, safety hats, gloves, safety glasses, overalls, earplugs and possibly raincoats and rubber boots. Often I've witnessed personnel arriving on projects and being unable to work for hours, or even days, due to there being no personal protective equipment available.

Insurances

The Project Manager must know what insurances the client has in place to cover the construction works. This includes, understanding the terms of the policy, what's covered and what's excluded, the duration of the policy, and what the excesses are on the policy.

The Project Manager needs to evaluate the risks and put appropriate insurance cover in place. It may be necessary to arrange additional insurance should the client's insurance be inadequate, or if the policy excesses are too high.

There are various insurances a project should always have, these include, amongst others:
- plant and equipment insurance
- insurance of the works
- public liability insurance
- workers compensation insurance
- professional indemnity insurance if there's design involved

The company may already have some of these policies in place making it unnecessary to take out additional cover. Nonetheless, the existing policies must be reviewed to ensure they adequately address the risks on the project. For instance, additional cover may be required if the project:
- is in another country, since some of the standard policies might not apply
- is located in a region prone to tropical diseases, where it's important to know personnel on the project will be covered in the event someone becomes ill
- is in a remote region, insurance should be in place to cover flying a patient home for emergency treatment, which could cost several thousand dollars
- contains particular risks which are specifically excluded from the policy

- has a value in excess of the existing policies
- requires the use of high value equipment items not included in the policies

If there are any doubts as to what insurance cover should be in place check with qualified experts on the matter such as, the insurance broker your company deals with.

To purchase additional cover the insurance broker will need to be advised of the type of project, its duration and any specific risks related to the project, such as constructing a bridge across a river or working in and around the general public or existing structures.

As happens when you insure your house, car or personnel effects, you can choose a variety of covers and the quoted cost will depend on the perceived risks and the excess you are prepared to accept. The Project Manager needs to evaluate the suitability of the various insurances, and I suggest the various quotes and options be discussed with senior management. Like any insurance you don't want to be part way through the project, have an accident and then find that the project is underinsured or not insured at all.

Bonds

With most projects the client will expect the contractor to supply a bond or surety to ensure the contractor will successfully complete the project. Normally the bond will be up to 5% of the contract value, half of which is released on substantial completion, with the balance released at the expiry of the warranty period, often twelve months or more after substantial completion. Bonds usually have to be in place before the client will allow the contractor to start work on the project. It's, therefore, essential that the Project Manager ensures the application for these bonds is made as soon as possible.

Clients normally want the bond supplied in a specific format with specific wording. Failure to comply with this wording may mean the client won't accept it, forcing the bond to be reissued. Often this wording is onerous and a bank may not want to use it, so it's sometimes necessary to negotiate the wording with the client.

Most clients usually require the bond to be issued by a bank. If the bond is issued by an insurance company it's often cheaper, and consideration should always be given to requesting the client accept one instead.

Since the bond is normally released in two portions, it's advisable that two separate bonds be obtained of equal value, their total sum being equal to the value of the bond required. In other words, if the contract requires a bond of 5% with 2½% to be released when the contract reaches substantial completion then get two bonds each to the value of 2½% with one issued for the expected date of substantial completion, and one for the expected date of the warranty period. I normally ask for the bonds to be issued for a month or two longer than expected because most contracts are delayed to some extent, and it's sometimes difficult to get the bond reissued or extended. (Some clients may not accept a bond with an expiry date.)

It should be noted here that if the contract value increases, due to variations being issued in the course of the contract, it's usually necessary to obtain a bond for this additional value.

Many clients not only request the contractor supply a bond, but also retain retention money from each payment made to the contractor. Again, this is normally

5%, but can be 10% with half released on substantial completion and the balance at the end of the warranty period. I would always question the client on the necessity of having retention money as well as a performance bond, especially after they have taken occupation of the facility. Some clients will accept a retention bond in lieu of holding retention money. Depending on the circumstances of the construction company, such as their cash flow position and available bond facilities, the duration of the contract and warranty period, will influence whether it's worth requesting the client accept one. The bond obviously has a cost to it, but the contractor loses out on the interest which the client earns from retaining retention money. The retention bond normally only has to be given to the client before the first payment is due from the client. Failure to issue a retention bond means the client will hold retention monies until such time as they receive the bond.

Sometimes the contractor requests the client issue a payment guarantee to ensure the work done by the contractor is paid for. However, most major clients will not be prepared to issue one. If the contractor requires a payment bond it's essential this requirement is incorporated into the tender submission, and that the bond is received before the contractor starts work on the project.

The payment guarantee must be checked to ensure that:
- it's issued by a valid and reputable banking institution
- it's made out in the correct name of the client that you have contracted with
- the wording on the bond is such that the contractor has rights, and is able, to call in the bond should payment not be received from the client
- there is no expiry date on the bond, or if there is, that this date is beyond when the final payment is due to the contractor, taking into account the retention the client may retain

Case study:

A number of years ago the company I worked for was constructing a golf course for a company who ran into financial difficulties part way through the project. Since we had a payment guarantee we were unconcerned and continued working assuming we would be able to call in this guarantee. When the client defaulted on our payments we attempted to call in the guarantee from the bank, only to find it was issued in the name of an entity different to the one we had contracted with, and we therefore had no entitlement to it. This was a costly mistake – over twenty million dollars.

I would recommend Project Managers personally check the payment bond and then request the legal and financial experts within the company to check it as well. An error with the wording can be expensive to a contractor should the client not pay for work completed. The bond should then be stored in a safe at the contractor's Head Office to avoid the risk of it getting lost in the project offices.

It should also be noted that if the contract grows in value, then it will be necessary for the client to issue a revised payment bond to take account of the new contract value.

Permits

All permits must be in place before starting the project works and the Project Manager must understand from the contract who's responsible for providing which permits. There may also be requirements for building and environmental permits. These are often, but not always, obtained by the client, even so, it's the contractor's responsibility to ensure that all permits are in place because work shouldn't start without them. Failure to have the correct permits can result in fines being issued for the breaches, and further work being halted. While permits not being issued could cause serious repercussions for the client, possibly even cancellation of the project.

The contractor must obtain excavation permits and those for constructing the temporary offices and stores on site.

Notifications

Prior to construction work starting there is often a requirement to notify various local council authorities. Even if there isn't a requirement it's possibly wise to notify them and other affected departments (like traffic, fire, police and ambulance), as well as any neighbours that may be affected by the project.

Licenses, registrations, certificates and qualifications

When the project is in another jurisdiction, particularly another state or in another country, the Project Manager must ensure that the company is registered with the relevant trade organisations, state and government departments (including for tax), and has the required licenses to undertake the work. In addition the Project Manager may be required to have particular qualifications, registrations and licenses to enable the contractor to carry out the work. Various trades may also require licenses and qualifications to be in compliance with the legal requirements in the project jurisdiction. Failure to comply with these requirements could result in the project being stopped, delayed, or the contractor and Project Manager being fined.

Some clients may also require the Project Manager, their staff, and workers to have particular qualifications or trade certificates.

Quality management plans

Prior to work starting, a project quality management plan should be prepared and submitted to the client for approval.

(This is discussed further in Chapter 9).

A word of warning here, often the Project Manager relies on the project Quality Manager, or even the quality department within the company, to draw up this plan, and then neither the Project Manager nor the rest of the project team reads or has any input into it. This can lead to quality plans being poorly 'cut and pasted' from previous projects. I've seen quality plans submitted to the client with the incorrect project name, and including tasks which are not part of the project. In addition, sometimes these plans are more complex than required for the project, slowing down the approval processes required to carry out the work. It's, therefore, important that the Project Manager reviews all quality plans and other quality documentation before submission.

Client deliverables

After award of the contract the client usually has a list of deliverables which need to be submitted within a specified time frame. These will include, amongst others, the contract schedule, safety plan, environmental plan, quality plan, method statements, insurance policies, bonds and guarantees. Some of these may have to be submitted before work can start on site. As a Project Manager goes through the project documentation they should prepare a list of these deliverables and allocate members of the project team to prepare them.

In some cases these plans will have to be approved by the client or their representative before work can proceed. Unfortunately in many cases the contract may not specify how long the approval process should be, and the onus will be on the contractor to get the approval in writing, as soon as possible.

Sometimes it can take several weeks for the client to work through the plans and then they often come back with comments that require the plan to be redone and resubmitted, which can delay the start of the works. In certain cases I've found, or suspected, that this is a delaying tactic by the client because they weren't ready with the site access or were unable to issue the construction drawings. So by rejecting the contractor's plans they can put the blame back on the contractor for causing a delay, thus avoiding the contractor lodging a claim for late access or late drawings.

It's therefore essential the Project Manager ensures the project plans and other deliverables are submitted as soon as possible, and certainly within the time frames stipulated in the contract. In addition, the documentation must comply with the requirements and wording in the document. If in doubt as to how the document should look, and the detail required, ask the client or their representative before starting work on the plan.

Project plan

Many clients require a formal project plan be prepared for the project, but even if this isn't a client requirement, it can be beneficial to prepare one. It should include:
- major tasks on the project
- an outline of how major tasks will be done
- equipment and materials required to do the tasks
- who will be responsible for the task
- key risks that could be encountered
- what actions will be implemented to mitigate risks

The purpose of the project plan is to ensure the project team has thought through the processes associated with the project and planned how these will be tackled to ensure they are accomplished safely, efficiently, to the required quality, and in accordance with the contract schedule.

The Project Manager should review the project plan, since often the preparation of it is delegated to inexperienced staff who 'cut and paste' plans from other projects that may not be relevant or appropriate. In addition, the project staff may not have thought through the processes required for the particular project.

Mobilisation check list

In preparing to mobilise to the project site I recommend a check list is used to ensure that all the requirements to start the project have been met, that the staff mobilised are able to start work immediately establishing the site facilities and offices, and that they will be able to begin construction work in accordance with the contract schedule.

Risk review and understanding the risks

In final preparation for the start of the project the Project Manager should prepare a risk review sheet for the project which should have the:
- risks itemised
- possible causes that could create the risk
- mitigating actions to prevent or reduce the risk
- estimate of the cost to the project should the risk turn into a reality

This schedule must be shared with the key project staff as well as company management. During the course of the project it should be reviewed at least bi-monthly to update mitigating actions in place, and assess if some risks have diminished or if new risks have become apparent.

Most projects also have opportunities, it would be prudent to prepare a list of possible opportunities which can be investigated and pursued in the course of the project.

Summary

- Before starting a construction project it's important that the Project Manager goes through a thorough planning process.
- They must read all the tender documentation, the tender submission, the contract documents and the drawings to gain an understanding of the details of the project.
- They must also understand what structures will be constructed, where the project is located, what the various client and contractor responsibilities are and any project specific conditions.
- Only then can the Project Manager plan how they will construct the project and what resources will be required for the construction.
- A schedule must be prepared for the project.
- Insurances and bonds must be put in place.
- Various documents (like safety and quality plans, method statements, project organisation charts and permits) required by the client must be submitted and approved.
- The Project Manager must ensure that offices and site facilities have been ordered and that there will be utilities and services in place to enable construction to proceed.
- Orders must be placed for the subcontractors, materials and equipment which will be used both for the setting up of the site facilities, as well as for the first stages of the construction works.

Chapter 2 - Starting the Project

Nearly as important as the planning phase of the project is ensuring the project starts correctly. How the project is begun often dictates the way it will run. Good relationships must be developed amongst the staff, with the client, the local community, as well as with the suppliers and subcontractors. The rules, standards and processes established at the beginning of the contract must be correct because they will probably be those that are followed during the course of the project. It's important the Project Manager gives extensive thought to these, and takes the lead in their implementation since it will make the running of the project much easier.

Many things happen in the first few weeks of the project and often Project Managers are so focused on reaching the first milestone (pouring the first concrete, erecting the first steel, installing the first items of equipment), that they overlook many of the basic planning structures required.

If everything has been planned correctly, and the management staff and workers are arriving in the correct numbers and sequence, start-up will be simpler. The manning requirements must allow for sufficient resources to do the site establishment, set-up and erect the project offices, stores and workshops, while at the same time starting construction work on the project.

Facilities

Priority should be given to set up at least one or more offices that will have power and internet connectivity. I've seen many projects where this has taken several weeks to occur, resulting in the project staff wasting much of their time going back and forth to use another company's facilities to receive and send emails, and copy documents. Without a proper office, documents aren't stored or filed properly, resulting in documents being mislaid, or the incorrect documents or drawings used. If project staff don't have a place where they can open up drawings and documents, and work from them, there is a chance that mistakes will be made.

In addition, toilet facilities must be operational at an early stage. Workers lose time walking or traveling far to use other facilities. Obviously if there are no toilets available it's unlikely the client will allow the construction to proceed. Lack of facilities can also lead to industrial relations problems.

What happens at the start of the project, and how efficiently the contractor builds their offices, and completes the establishment of their facilities, is keenly watched by the client and managing contractor since it often provides a clue as to the quality and professionalism the contractor will give to the construction of the permanent facilities.

The layout of the project site facilities must be planned and approved by the client and the necessary regulatory bodies. Layouts should take into account any future expansions that may be required to the offices, how the site facilities will be delivered and offloaded, the size and turning circles of delivery vehicles, as well as the equipment required for offloading the units. As important, is planning how the units will be dismantled and removed from the site when construction is complete, especially since other contractors may be working on the project at this stage which could impact on the process. I've heard of contractors having to hire large cranes to dismantle their office facilities due to there no longer being access to get the units loaded on to trucks.

The site facilities should be set up in a professional manner and the offices should be of a reasonable quality since this is what the client will see every time they visit your office. Badly set up offices that are in a poor condition don't create a good impression. Since project staff will be working in these offices for the next few months, and sometimes years, the offices should provide a reasonable degree of comfort and space, and be aligned neatly and set level. It takes as much time and effort to set them straight, square and level, as it does to put them up badly. It should also be remembered that this is the first test for the construction team, so they should get used to delivering quality work from the start.

Setting up the safety systems

(Refer also to Chapter 5 for more on safety)

Case study:

A project I was asked to take responsibility for had been under construction for two months and had several months left to run before completion. The safety on the project was generally poor with incidents and accidents almost daily. These incidents, although minor, were taking up a lot of management time because each time they happened the person had to be taken for medical treatment, and often the injured person was unable to immediately return to normal duties. Each incident also had to be investigated, taking more management time away from their duties.

The main problem was nobody on the project took safety seriously. No-one took ownership of it. The safety standards were never clearly defined, workers weren't properly inducted onto the project, nor were there safety audits done.

When it comes to dealing with safety, quality, human resources and the client, it's always difficult to change processes part way through the project. Human nature is strange; if rules are set at the beginning, and a particular standard is demanded then it will be relatively easy to implement and maintain the system. However, if a low standard is set at the beginning, then part way through more stringent rules and requirements are put in place, they will be difficult to implement because people have become used to working at the lower level, set in their ways, and reluctant to change to the new rules and standards.

Tasks to be arranged include:
- Set up all safety documentation in accordance with the project's safety plan.
- Establish registers, (for power tools, slings, scaffolding, excavations, accidents and so on).
- Organise files for all the safety documentation.
- Formally appoint the designated responsible persons, ensuring they have the correct qualifications and experience, and that they are aware of their duties and responsibilities.
- All workers must be issued with, and use the required personal protective equipment.

- Install fire extinguishers.
- Make first-aid kits available.
- Display clear safety signage and posters.
- Inspect equipment and tag as safe.
- Ensure all equipment brought on from other projects and belonging to personnel is compliant (I've often encountered cases where Supervisors have brought their tools from another project that don't comply with the safety standards, and don't have the project compliant tags on them. Yet they end up being used simply because they were available and no one bothered to check them.)
- Hazard assessments are in place for all tasks.
- Materials are stacked correctly and in an orderly fashion, and dangerous and flammable liquids have their own ventilated stores with bunds to contain spillages.
- Plant and equipment arriving on the project is inspected to ensure it has the required paperwork and service history, and that it complies with the project requirements and safety standards.
- A training register is compiled, and as new personnel arrive on the project their details are added to the register. The register must have the name of the person, details of their qualifications, types of licenses and the expiry dates of the licenses. This register must be maintained during the course of the project to ensure that licenses are always current and that personnel are competent to undertake the tasks they are performing.

Welcome packs

It's good practice to issue a 'welcome pack' to all workers who will be transferred to the project. This is a document that gives the employees information of where they'll be going, and a brief overview of the project. These packs are especially important for workers travelling a distance from their home towns who may not be familiar with local area rules, customs or currency, or when projects are in remote areas with limited infrastructure and little to no mobile telephone coverage.

It should include:
- the location of the project
- climatic conditions of the area (temperature and rainfall)
- accommodation and transport arrangements
- specific project rules
- contact details for the project, including postal address and telephone list (remember most employees will have family at home who would like to know how to contact their loved ones)
- what tools and personal effects they should take
- the expected overall duration for the project
- formalities workers can expect once they arrive on the project (For instance, they may be required to attend a client's induction, undergo a medical examination and attend a company site specific induction. Many workers arrive on a project uninformed of the process involved to start

work on the project, leading to frustration when these processes appear to take a long time.)
Sometimes it's useful to include a few photographs of the project site.

Project induction

All personnel employed on the project, including subcontractors and staff, should undergo a company project specific induction. Often personnel may have undergone a client project specific induction, but this doesn't cover the specific works the company will be carrying out, the contractor's company specific rules or the specific risks associated with their work on the project.

Ideally the induction should be held in a room with seating and ventilation, and should be presented as a Power Point display. Obviously at the start of a project this is not always possible, however it's essential that all employees go through some sort of induction, so a printed copy of the topics might have to be sufficient until proper facilities are established. Once offices are established and it's possible to run the full Power Point induction, it would be valuable to re-induct everyone again.

I've attended many site inductions, both the client's and our own, and in general found them of a poor quality, and only focussed on general safety rules and regulations.

Conducting the inductions is often left to the safety department and results in the inductions sometimes being meaningless to most of the personnel. The Project Manager, or the next most senior manager, should at least give the welcome and a brief introduction. This will give the induction importance, as well as making the new employees feel welcome. I would also encourage the Project Manager to play a role in preparing the induction so as to ensure that the facts presented are correct, relevant and cover all the pertinent points.

The induction should include:
- a welcome address
- a brief overview of the project, which could include who the client is, a description of the overall project and the specific structures you will be constructing
- the duration of the overall project and the expected duration for specific tasks
- an update on the status of the project, the milestones met to date, and milestones due in the near future
- specific problems encountered on the works, or that may be encountered
- possible other works on the project, which the company may become involved in in the future
- a section giving a brief overview of the company (if there are new employees, or subcontractor employees)
- the company values
- a general induction focussed on the workers themselves (personnel are often focussed on their personal problems and questions, so once these are addressed they will be more focussed on the content of the rest of the induction), which covers:
    - work hours
    - accommodation

- daily transport
- location of toilets, offices, stores and lunch room facilities
- project rest days
- dates when the workers can expect to return home for their rest break and the length of this break
- transport arrangements for the rest break
- the overall site management structure, mentioning people by name and their individual responsibilities
- project site specific rules such as:
  - areas that personnel must not enter
  - site access routes
  - location of services
  - location of all emergency equipment, including a plan of the project site
  - rules for the workers accommodation
- company disciplinary and grievance procedures, including which offences will result in workers being removed from the project or dismissed
- safety (this shouldn't just be a discussion on general safety topics but should focus on the actual project specific safety rules and hazards)
- environmental issues which should include:
  - site boundaries
  - environmentally sensitive areas
  - local fauna and flora that may be encountered
  - procedures should local fauna be encountered
  - disposing and segregating of waste
  - actions in the event of an accidental spill
- procedures to follow if a worker becomes ill or injured, either during work hours or after work
- emergency contact details (a useful practice is for employees to be given a small laminated card listing emergency contact details, which can fit into a pocket to be carried with them at all times)
- the quality expectations for the project

At all stages of the induction employees should be encouraged to ask questions.

The person must also be made aware of where they should direct their specific problems or complaints. I often find that employees don't know who they should report problems to concerning pay or leave queries, so they either don't report them, which leads to them being upset, affecting their morale and work performance, or they report them to a person not responsible for resolving the problem, which often means the problem isn't addressed, causing frustration with the worker and again affecting work performance. On numerous occasions when I've visited a site as the Project Director I've been approached by workers reporting issues relating to short pay, hours of work or other complaints. In many of these instances the Project Manager is unaware of the problems because the employees have either not reported the problem, or have reported it to the wrong person.

Lastly, if possible, new employees should be introduced to the person they'll be reporting to.

Everyone who attends the induction must sign an attendance sheet which must be kept with the safety records. I would also consider setting a brief questionnaire for the participants to confirm they have heard and understood the induction.

Don't forget to ensure that the induction is updated regularly because progress and rules change as the project progresses, as may staff.

Staff induction and welcome

After staff have attended the company project induction it's important that the Project Manager, or person they are reporting to, goes through the rules, requirements and expectations on the site.

This should focus on:
- what their role will be on the project
- who they will be reporting to
- who will be reporting to them
- the lines of communication
- their specific areas of responsibility
- their levels of authority
- their responsibility regarding safety
- their responsibility in terms of quality
- documentation they must prepare and submit, and the deadlines for this
- the limits of their authority
- what problems and issues should be referred to the person they report to, and what problems and issues they're expected to resolve themselves

A lack of understanding of what the Project Manager and the company expects from individuals often results in poor performance, to say nothing of the importance that all staff are fully aware of their responsibilities.

Staff should also be aware of the authorisation processes on the project as well as within the company. These limits relate to purchase and subcontract orders, processing of time sheets, tenders, variations, and claims.

Furthermore staff must know who to contact in the various departments within the company when assistance is required.

A project organisational chart, as well as a company organisational chart, which clearly defines the organisational structures and channels of communication will help clarify the various roles and responsibilities.

Site hours

The Project Manager and their team must decide at the start of the project what the work hours are for the project. These are often dictated by the hours worked on other company projects, union or labour agreements in place, or by the project specific rules. It's important these agreements are read by the Project Manager, and are understood and adhered to.

Be aware that there may be industrial relations problems if the work hours set at the start of the project are reduced half way through the project. Most workers are happy to work long days since they are normally paid hourly rates with the possibility of earning overtime. They can become used to receiving an income every week based on these longer hours and are therefore unhappy when their pay is reduced when the project cuts back on the work hours.

In general, I've found that it's counterproductive to employ manual workers for more than ten hours in a day. This is even more so when the project is located in an area of extreme temperatures. Also consider the workers' travel time to and from their accommodation, which can be in excess of an hour each way, resulting in workers having twelve hour work days or longer, which in turn leads to poor productivity and compromised safety.

The site work hours should take into account the hours of daylight, since work should preferably be done while it's still light, unless there's a requirement to work at night. Starting times may be varied to allow for an earlier start time in summer so as to enable as much work to be done during the cooler hours of the day.

Accommodation for personnel

The responsibility of providing accommodation for personnel (when required), can either be the client's or the contractor's, depending on the project conditions.

If it's supplied by the client this takes some of the responsibility away from the contractor, however they must still satisfy themselves that the accommodation provided is suitable, meets the requirements of the current labour agreements, is safe and secure, is kept in a clean and hygienic state and that maintenance is done regularly. The Project Manager should consider establishing an accommodation committee which can ensure that any relevant problems are reported and attended to quickly, before they become an industrial relations problem.

Personnel should be given, and sign receipt for, the accommodation rules. These rules must be enforced from the start of the project. When a person moves into the accommodation an inspection should be done and defects recorded. Then, when the person moves out, the accommodation should be reinspected to ensure there has been no damage. If there is it may be necessary to hold a disciplinary hearing with the person.

If the contractor has supplied the accommodation it should:
- comply with safety regulations and have fire extinguishers and first-aid facilities provided
- have safety posters and emergency contact details prominently displayed
- be kept clean and hygienic
- have adequate provision for the disposal of sewage and general rubbish
- be regularly maintained

The Project Manager ought to check this accommodation regularly to ensure it is being cleaned and maintained, and is in a safe condition. All electrical appliances used by the occupants must be safe to ensure they do not cause fires, electrocute people, or overload the electrical systems.

Transport of personnel

In towns and cities workers normally have to make their own transport arrangements, although sometimes they might have to be recompensed the cost of this. It's important the Project Manager understands the requirements for this reimbursement, which is normally outlined in the project labour agreement, the company labour agreement, or in a union agreement.

Many projects, particularly those in remote areas or country towns, where the contractor is required to provide accommodation for the workers they will also usually have to provide transport for the workers to and from work as well. Although, occasionally the client is responsible for providing the accommodation and the transport, so again the Project Manager must understand the contract with regards to who is responsible.

The transport provided must:
- be safe
- in compliance with the local transport and traffic regulations
- comply with any project specific safety requirements
- have a driver with the correct valid license, who doesn't drive recklessly, break road traffic rules or project rules, or drive under the influence of alcohol or drugs
- not be overcrowded
- make provision for workers who have to work shifts or extended hours
- have adequate insurance in place should there be an accident

Often third party transport companies are used to transport workers, however I've found in the past that if there is an accident, workers may not be covered by insurance.

Case study:

A contractor's bus transporting their workers was involved in a serious accident, leaving several workers severely injured. Resulting in them being evacuated by medical air ambulance, spending several weeks in intensive care, incurring other medical treatment costs, plus compensation for the time the workers were unable to work. This amounted to several hundred thousand dollars which were not covered by their insurance company because the project was in another country and the transport was provided by a third party.

The transport provided must also be reliable. The failure of a bus to promptly collect workers from the accommodation can result in unproductive time since the workers will have to be paid for the time lost because they weren't responsible for the delay. Just as serious is the failure of the transport to arrive on time on completion of the shift, since this can lead to industrial problems with the workforce, and the project having to pay the workers for the additional time while waiting for the transport.

When the transport is planned the Project Manager must take into account the conditions in the subcontractors' documents, and who is responsible to transport their workers to and from site. If it's the subcontractors' responsibility it may become necessary to monitor overcrowding, and who uses the transport, to avoid the subcontractors' employees using the main contractor's transport (even though the subcontractor is responsible to transport their workers).

Check all transport vehicles regularly to ensure the vehicles are maintained, and kept in a clean and safe condition.

Site fencing and security

The quality, level and type of fencing and security on a project are dictated by numerous factors, including the client's requirements, the location of the project, existing security and fencing around the project and the level of security risk, as well as the likelihood of a member of the public having unrestricted access to the project. Local councils often have bylaws which stipulate the type of fencing or hoardings.

Most contract documents will include the security requirements for the project site. If there's no clarity it would be prudent to ask the client what they will be doing to secure the project site and encourage them to put these measures in place as early as possible. Note, however, that even if the client has employed security guards and put perimeter security fencing in place, each contractor will still be responsible to protect their own equipment, facilities and works.

If the site security fencing is in the scope of the contract consideration should be given to erecting this at an early stage of the project. Not that this is always possible since often the fence cannot be constructed until other works on site have been completed.

Entrances to the site should be controlled so that no unauthorised visitors can enter, thereby creating a safety hazard since they normally wouldn't have the appropriate safety equipment, haven't been inducted onto the project, and would be unaware of the site's risks and hazards. Unauthorised visitors can also damage property and possibly even steal from the site.

Some projects are spread over many kilometres, for example roads, railways or pipelines. These projects are almost impossible to fence completely. However, it will be the contractor's responsibility to ensure that members of the public are kept out of areas of risk. This could be done with signage and temporary barricading around deep excavations and other hazards. Locations of offices and areas where the equipment will be parked overnight should have security fencing around them or security guards for after-hour periods.

Even if the likelihood of theft is minimal I would recommend that some form of security is provided. After all even a minor theft, like a battery from an excavator, can result in a major cost to the project, especially on a remote site where it may take several hours or even days to procure a replacement battery, resulting in the excavator being inoperable, as well as the trucks the excavator was loading standing idle. In some areas, thieves target copper, and in doing so can severely damage an electrical distribution board for a few dollars of copper. The cost of replacing the wire is small, yet repairing, or replacing the board could be a couple of thousand dollars, but more significant is the fact that it could take several weeks, or months, to manufacture and resupply the replacement board. Which could, of course impact the project schedule and result in a key milestone being missed.

The Project Manager must understand the security risks of working on the project site and take appropriate actions with due regard to the risks and implications of property theft. In some countries where property theft is committed by armed gangs, it may lead to injury or death of personnel. In these areas I've been forced to install perimeter fencing with armed security guards and a backup response unit.

Apart from security fencing and the fencing required to restrict unauthorised visitors, the project may have other fencing requirements. Perhaps it contains areas of heritage or environmental value, in which case it's good practice to fence these areas with a simple fence and appropriate signage, to prevent any personnel accidently entering the area.

Alternatively, the project site may be in a farming area making it necessary to keep livestock out of the work area. After all, it wouldn't be good for the local community relations to have a cow fall into an excavation on the project. Remember that livestock cannot read signage, so fences must be strong enough to prevent animals from entering the site by force, or by going under, over, or around the fence.

Access to site and access on site

Maintaining the access roads to the site, and on the site, during the construction phase, is normally the contractor's responsibility. This may extend from actually constructing the roads (which may or may not be part of the contractor's project scope for the permanent facility), to maintaining the existing roads and ensuring that they are handed over to the client at the end of the project in a similar condition to their state at the start.

On remote sites, or sites covering a large area, the construction of access roads can be a significant cost, and if the client has not provided these roads it will be the contractor's responsibility to build and pay for them. There's always a temptation to build the roads as cheaply as possible, but poorly planned and built roads may:

- lead to them breaking up during the project's construction adding additional repair costs and possibly disrupting the work
- jeopardise productivity on the project, particularly with earth moving projects where trucks hauling material cannot travel at the maximum production speed due to the poor quality of the roads
- result in damage to vehicles, particularly to tyres, which not only cost money to repair, but results in the vehicle or machine being out of production while the repairs are performed
- be a safety hazard

It's important the roads are safe to drive on. This includes appropriate speed restriction signs, as well as warning signs and Stop signs. Where the site access road meets the public road there must be clear lines of sight for vehicles travelling in all directions, and if necessary provide additional turning lanes.

It should be noted here that when any work is done on a public road, or a new access road is built entering a public road, appropriate permits and planning permission will be required from the authority that is responsible for the public road.

Safe access must be provided for personnel on the project.
- Walkways should be separated from vehicle traffic.
- Personnel must be directed away from the edges of excavations and drop off points.
- Safe access must be maintained into excavations and on to elevated structures.
- Access must be provided between the site facilities and work areas.
- Access must be checked and maintained during construction since access-ways may have to be moved to accommodate activities and new structures.
- Access routes must be clearly marked.
- Personnel must be encouraged to use the access-ways, since there will always be the natural inclination to take short cuts across the site, which may then endanger the person.

Parking areas

Sufficient parking should be allowed where possible on the project, however it can be costly and on some sites there may not be an area available. This parking could have to take into account employees' private vehicles if they are required to use them to travel to site, although where possible employees should be discouraged from using their private vehicles. In some areas this parking might have to be secure to prevent possible vandalism or theft of vehicles.

Sometimes it's a requirement to provide sufficient parking for delivery vehicles arriving at the project entrance gate thereby enabling them to park safely, without blocking and interrupting other road users, while the drivers are signing in or waiting for project staff to direct them to the offloading area. This can be a particular problem encountered for deliveries entering high security areas or for multiple deliveries, like concrete trucks that have to wait for the delivery ahead of them to be discharged before they can enter the project site.

On the project site there must be sufficient parking set aside for the peak number of construction vehicles and equipment, both during the day and at the close of the shift. Congested and poorly planned parking areas can lead to vehicles accidently coming into contact with each other. In addition, if the parking area is congested there'll be delays in moving the vehicles at the start of the shift which will impact on the productivity.

Diversions, deviations and road closures

On some projects it's necessary to divert traffic on to alternative roads to enable work to be performed on the existing road. This will require special permits and authorisations from the relevant authorities governing the road, as well as from the local traffic authorities. These deviations will have to be done in conjunction with the client. Safety is paramount in these situations and the correct procedures and road signage will have to be in place. The closures and deviations need to be done with the minimum disruption to existing traffic. Advance warning issued to the people using the roads will help with this, while also making them aware of the changes. This notification may be in the form of press and radio advertisements or letter drops to local residents or businesses using the route.

Diversions and deviations can take several months to plan, and receive the required permits and authorisations. The Project Manager should organise these long before they are required to be actioned. Local residents and businesses affected by these diversions must be incorporated into the process and should be notified well in advance so they can make alternative arrangements if required.

All the required barricading and signage must be installed in accordance with the local by-laws and traffic ordinances. The signage must be clear, unambiguous and must be visible after dark. Lighting may be required to illuminate the general area (ensure the lights are angled correctly so the glare does not impede vision), as well as warning lights and directional signs. The barricading and lighting will have to be checked frequently to ensure that it hasn't been damaged or moved, and is still relevant to the progress of the work.

Sometimes pedestrian traffic also has to be diverted away from and around the project site, and again, this must be planned correctly, and communicated to the neighbours in advance.

In cities where projects can have limited space for offloading of deliveries it could be necessary to arrange these deliveries for after-hours, or to close part of the public road to allow the delivery vehicles to park. Once again, the project will require authorisation, and people affected by the closures will have to be forewarned.

Stormwater

Stormwater should be considered when planning the laydown areas and access routes since even minor rainfall can cause flooding if there's no adequate drainage. Inadequate planning and handling of the stormwater can lead to environmental problems caused by erosion of the site and surrounding properties, and the deposition of silt on the project site and surrounding properties.

Apart from environmental issues there's the risk of localised flooding, causing damage to buildings and equipment, on the site or surrounding properties. When this happens the excess water must be pumped from the areas at additional costs, adding the potential problem of where it should be pumped to, and work being unable to continue until the water has been removed.

If the installation of the permanent stormwater drainage system is part of the contract, consideration should be given to completing these works as soon as possible. However the new stormwater drains often connect into structures or systems that are not part of the contract, making it prudent to ensure the client or the authorities responsible for the downstream work complete the works before the project's stormwater system is completed. After all, if the downstream works aren't completed it's still likely the project site will flood.

Once the stormwater system is completed it will have to be maintained. Where possible, measures must be put in place to ensure that silt and rubbish doesn't enter the system causing blockages and requiring the system to be cleaned out before the project is handed over.

If the stormwater system is not part of the contract the client should be encouraged to install the permanent system as soon as possible.

As construction works progress the landscape of the site will change. These changes could be due to the delivery of material, construction of temporary or permanent structures, new excavations, or new roads. Even insignificant changes can alter the characteristics of the stormwater management on site with drastic results. Therefore it would be good practice before each expected rain event, for a senior member of the construction team to walk the project site with the sole aim of assessing the stormwater management systems in place and ensuring that they aren't obstructed or requiring maintenance. After each storm it would also be wise to assess the storm damage and ensure the system has worked, and, if necessary, implement any repairs or modifications to improve the system.

Where possible all excavations should have temporary berms formed around them, plus drains constructed on the surface to lead the stormwater away, thereby preventing any water, other than direct rain water, from entering the excavation. Water entering the excavations not only damages property and equipment in them, but will often destabilise the sides of the excavation, resulting in an unsafe working condition, which will have to be rectified before further work can proceed.

Signage and posters

Various signs have to be provided and placed on the project, including those advertising the construction company. Authorisation for these will be required from the client, and possibly also the local and traffic authorities. Since both the client and public will see them they should be professional and erected in a manner that will reflect the qualities of the company. The sign must be fixed in such a way that it will not be a safety hazard, obstruct visibility, cause a traffic hazard or become damaged in severe weather conditions.

There may also be a requirement for a project sign board, the details of which would normally be in the contract document. It's important before the sign is created that the manufacturer provides a template of it so the Project Manager can check to ensure the wording, spelling and company names are correct. I've had, on more than one occasion, the embarrassment of receiving a project sign on site with incorrect spelling, and then incurring the cost and delay of having to send the sign back to the supplier to have the errors corrected.

The foundations and supporting structures for the signs must be designed to withstand the worst weather conditions that can be expected on the site. I've seen many project sign boards flattened by the first strong winds, which isn't a good advertisement for the contractor.

In addition to company and project signs there are often requirements for directional signs. These are normally provided on the site access road, and within the site to guide deliveries and visitors to the project offices. On large sites it could be advisable to divide the site into different designated areas that are clearly marked with direction arrows and boards. This will enable deliveries to be directed to the specific areas where the material is required.

Safety signage is essential. These signs indicate:
- the location of first-aid boxes, muster points, emergency wash stations, spill kits and firefighting equipment
- hazards on the site
- road signs warning of hazards, speed limits and to stop or caution traffic
- the personnel protective equipment to be worn

Failure to have the correct and adequate safety signage on the project can lead to unsafe situations, and many clients won't allow work to proceed without the correct signage. All safety signs must be manufactured in accordance with the regulations, be prominently displayed, and securely fixed so that they don't fall down.

Where personnel enter the site I usually place a project notice board, which displays the company's policies and procedures, emergency procedures, emergency contact details, details of the safety representatives and employees trained in first aid, the safety meeting minutes, minutes from the latest toolbox meetings, and other relevant safety bulletins. The project notice board should be updated on a regular basis to ensure the information is always correct and relevant.

At the site entrance it's also good practice to include a board with the project's safety statistics. This would include the total hours worked on the project to date, the hours worked since the last safety incident, the number of safety incidents recorded to date, and the number of lost time injuries recorded on the project. This board should be updated on a daily basis.

Company policies and procedures must be displayed on the project notice board, in offices and in the worker facilities. Relevant safety information, bulletins and notices must also be displayed in these areas, and updated on a regular basis.

Lighting

Lighting is required on site for security purposes, as well as to enable work to be done safely after dark, and to light up work areas which may be hazardous to other contractors or members of the public that may have to use the facilities at night.

Lighting can be expensive and many projects don't allow adequate funding. Most projects don't plan to work at night, but often circumstance (like equipment breakdowns or late deliveries of materials), require project personnel to perform tasks outside normal working hours. It's best to plan for these events well in advance and at least have permanent lighting situated around the office areas, main items of equipment, and strategic locations around the project. It should be remembered that project staff often work extended hours in the site office to complete unfinished paperwork from the day. If these areas are dark it can result in staff injuring themselves falling over obstacles.

When a task is performed during the hours of darkness there must be sufficient lighting in the task area, as well as around the offices, stores, ablutions, entry and exit points to the site, and along the routes connecting these areas. Again, lack of sufficient lighting in these areas can result in injuries to the workers.

Positioning of lighting must be done in such a way that it leaves no areas of dark shadows, preferably, with the lights being placed high enough, so they don't shine into peoples' eyes. Cognisance must be taken of the fact that the project site will change as work progresses and in the future new structures (both temporary and permanent) may block some of the light.

There will also be a requirement for temporary lights that can be moved around the project site and set-up at task areas as required. These must:
- be brought onto the project site well before they are required
- have stands that are sturdy and safe, with a suitable base that will prevent the light from falling over in windy conditions
- have sufficient spare light bulbs
- be checked often to ensure they are serviceable

A project staff member should be allocated to check the lights are working and that the project has sufficient lighting during the hours of darkness.

As buildings and structures are completed and closed up natural light may be blocked from entering. It's then essential that adequate lighting is provided to enable safe access within the buildings to the different work areas and access points, and to ensure that the quality of the finishing works can be maintained, and work can be done safely.

Stationery

The Project Manager and team should ensure the appropriate stationery is available on site for the running of the project. This will include items like daily diaries, day-works sheets, request for information books, time sheets, material received sheets, site instructions, letterheads and material transfer sheets.

In addition, the project will require various stationery items like: pens, calculation pads, scale rules, erasers, date stamps to mark when drawings and correspondence are received and stamps to mark drawings superseded.

Document control and filing systems

The proper control of documentation is a vital part of any construction project because lost or mislaid documents can result in costly mistakes. The document control system must adequately deal with the distribution, storage, retrieval and delivery of the complete documents, in a timely manner and so that the distribution can be tracked. The system should be set up correctly at the start, and maintained during the life of the project.

There are many forms of electronic document handling systems, some may be specified by the client, or perhaps the contractor has their own system. These systems can be useful, however personnel using them should be trained and authorised to use them. The system must be able to distribute, action, file and track the documents.

Correspondence to the client, suppliers and subcontractors should be allocated a correspondence number. From the start, sequences of numbers should be allocated to different addressees and different types of correspondence. These numbers should be used in sequence to enable the correspondence to be controlled and tracked. All documents should be stamped with the date, and even time, of receipt.

On larger projects it may be necessary to allocate a person to look after the document control.

Correspondence registers should be set up and maintained for the duration of the project. Staff should be allocated to track the progress of documentation to ensure there are no delays due to documentation going astray, or not being returned from review within the time periods specified in the contract. The reporting system should be updated and used at client and subcontractor meetings so delays can be monitored and resolved.

Once the document control and filing systems are set up properly it will be much simpler and quicker to locate documents throughout the course of the contract. It will also result in less misfiling which can lead to wasted time, or worse, when documents are misplaced and not actioned, causing project delays, extra costs and embarrassment.

Documents and drawings should be formatted using standard templates that are distributed to all staff involved with documentation, as well as to subcontractors.

Electronic documents should be backed up at a separate location. The back-up should be done daily, or at the most weekly. Computers crash, are stolen and offices may burn down, leading to loss of data, which can have expensive consequences for the contractor. Another advantage of backing up the data from personal computers is if a person leaves the project, the data accumulated by that person is available to other members of the project team.

Consideration could also be given to providing fireproof storage safes for important documentation.

Preconstruction survey and photographs

On projects that involve working in, around, or in close proximity to existing structures, facilities and roads, or when existing facilities are used, it's good practice to do a pre-construction survey of the existing infrastructure. Most clients, neighbours, neighbouring businesses and local authorities expect that any structures and facilities that may be affected by construction will be returned to them in the same condition, or better, once the new project is completed. Unfortunately, many of these people will not remember the condition the structures and facilities were in at the start of construction, and what memory they do have is often overinflated. If the contractor doesn't want to repair damage unrelated to the construction works then it's a good idea to prepare a detailed survey report, including photographs, of all the structures, facilities and roads in the vicinity of the project. This report should be handed to the client, neighbours or local authorities impacted by the project, so that they can verify the condition of their structures before work begins.

Blasting activities and large earthmoving equipment in particular can have an impact on structures several kilometres from the project site. Therefore, for projects that involve this type of work neighbouring structures should be checked, surveyed and photographed for signs of cracks before the project activities begin.

Accepting handover of work areas

Before work is started in an area, the contractor must verify that the area is in accordance with the contract information and Construction drawings. Often the client hands over a completed terrace, or excavation, for the contractor to construct on, and the contractor must check that it's of the correct dimensions and levels. Sometimes the project involves fitting steelwork, piping or electrical cables to structures constructed by other contractors and it's important to check these structures are in the correct position and of the correct height.

Also check that the structures handed over do not pose a safety risk to your work. This is particularly important when taking occupation of excavations done by others.

There should be a formal document to record the handover, with the date, relevant survey data, and if possible attached photographs. Any exceptions and deviations from the contract, or drawings, must be recorded in the handover.

Excavation permits, and location and protection of existing services

The Project Manager must ensure that all services are clearly marked and protected, and that all personnel are aware of their location. Where the exact location of services is not known it will be necessary to locate them using various detection equipment and carefully potholing in the vicinity.

All projects will require excavation permits to be obtained before excavation work can be undertaken, and job hazard assessments must take into account any services which may be affected by the task.

Survey and setting-out

An essential part of every construction project is ensuring all new structures are constructed in the correct location. Normally the client would provide the setting-out information for the structures, either relative to an existing structure or in the form of co-ordinates. This information should also include the elevation, related either to an existing structure or a datum elevation. It's then the contractor's responsibility to use this information to set-out the new structures.

It's important that the setting out information is checked before work starts. On one project we were provided coordinates for a structure, yet when the Surveyor came to set the structure out we found them to be incorrect, putting the proposed structure several kilometres outside the project boundary. Sometimes the error isn't as obvious, so the contractor shouldn't accept the setting-out information without doing some basic checks to verify the structure will fit in, and relate to other existing or new structures, both in position and height.

When a structure is set out by a Surveyor the setting-out information must be clearly conveyed to the Supervisor and the workers constructing the structure.

Case study:

During the construction of a casino I was the Section Manager responsible for the games area, which included a bingo hall. Due to the project being fast tracked the finishes were fabricated in conjunction with the building construction, rather than the norm of being fabricated after site measurements were taken of the completed building work. The construction of the hall was nearing completion and the finishes were being installed when we realised there was a problem – the wood panelling and mirrors did not fit on the walls. Several hours later the Supervisor came to me rather sheepishly and admitted he had made an error when constructing the brick walls. He had set marks on the floor to indicate the inside of the brick walls, yet when the bricklayers started the brickwork they assumed the marks were set-out for the outside of the wall. Consequently the hall had been built with the external walls located 220 millimetres inside of where they should have been, meaning the hall was 440 millimetres smaller than it was designed. This major error meant the client had space for fewer seats, and that many of the finishes had to be remanufactured to fit the smaller room. The only upside for me was that the mirrors, which were fairly ornate, couldn't be used on the project, and were handed out to friends and family. Nearly 20 years later we still have a mirror in our house, a reminder of that basic setting-out error.

This illustrates how vital it is that the person setting-out the works clearly conveys the information to the personnel building the structure so they know what the marks and points are for. Ideally a sketch should be given to the Supervisor, conveying the information, so there can be no doubt whether the points are the centre, inside, or outside of the walls or columns, or if they are the drawing grid lines.

The client

The Project Manager must establish a working relationship with the client, the client's representative and the staff that will be interfacing with these people. Channels of communication must be set up between the various members of the team so that problems and issues can be attended to in a timely manner. For this reason consider arranging an informal get-together so everyone can meet. Having said this, remember any contractual issues must be kept on a formal basis.

Establishing relationships

It's good practice to establish cordial working relations with neighbours and the surrounding community. This can be encouraged by keeping the neighbours and communities informed of what the project is, what's involved, how they'll be affected (noise, road closures, dust, and so on), and the start and end dates of the various phases of the project.

Nobody wants to receive complaints from neighbours so most of us provide people with as few contact details as possible. Unfortunately, what then often happens is if the aggrieved person cannot contact you to lodge a complaint, they track down the client, or someone at your Head Office, who will invariably be your manager, leaving you with not only an aggrieved member of the public but an upset client or manager too. In the long run it's far simpler for the neighbours and members of the public to have your details and to contact you directly if they have any queries, concerns or problems.

Establishing good relations with other contractors on the project is also useful. Therefore, where possible, providing they don't compromise your work, other contractor's needs and access requirements should be accommodated. After all at some stage in the course of the project it may be necessary for you to ask these same contractors for assistance.

Quality

Quality systems and procedures must be put in place from the start of the project, and be in accordance with the project requirements and specifications. All members of the project team must be aware of these requirements and everyone working on the project must be accountable for achieving the desired quality standards. Substandard quality cannot be tolerated, and work that doesn't comply may have to be redone. It's up to the Project Manager to ensure all project staff are properly trained and equipped to achieve the required quality standards.

(See Chapter 9 to read more about quality management.)

Summary

The Project Manager must ensure the project gets off to a good start. Personnel arriving on the project should be prepared for the site conditions and the applicable rules, and this is done by issuing them with a welcome pack, as well as ensuring that they all attend a project specific induction.

It's important that from the start the project is set-up to conform to the quality, safety and industrial relations rules and requirements. There is much work to be done to set-up the project correctly which includes:
- constructing the project offices and facilities
- providing the services and utilities
- constructing site access roads
- providing adequate drainage

- installing lighting
- installing security fencing and setting-up security
- installing signage
- arranging accommodation
- providing transport
- setting-up safety systems
- putting in place an effective document control system
- locating and protecting existing services and obtaining excavation permits
- surveying existing structures and ground surfaces and notifying the client in writing of any defects or discrepancies
- implementing survey controls
- establishing good relations with the client, neighbours and other contractors

Chapter 3 - Scheduling (Programming) the Project

Case study:
A few years ago I joined a new company and inherited a project that had been running for two months. The project was the civil work for a new gas power station.

The actual site covered only a small area of around five thousand square metres, with the main structures consisting of concrete foundations for the gas turbines, and numerous smaller concrete structures to support other equipment around them. These structures required a couple of thousand cubic metres of concrete. There were also several kilometres of underground services, consisting of stormwater, oily water and dirty water drainage, potable water, electrical conduits, fire water and sewer. These services were all at different depths below ground, some more than three metres deep, and not only did they cross over each other but they connected to different structures. In some cases the services even went under the concrete structures which were founded at different levels, most between one and two metres below ground level, but one particular structure was nearly five metres deep.

At face value the contract schedule looked good and was detailed with several thousand activities. All activities were linked (although we later found the links were incorrect), their durations appeared realistic, there was a critical path, and the schedule showed the construction meeting the project milestone dates. However there were already problems when I took over the running of the project, and it didn't take long before I realised the schedule had major shortfalls.

The schedule viewed the different structures, and different services, in isolation and scheduled them each as if they were separate stand-alone entities. Consequently, for example, a delay to one of the services caused no delay to the structures on the schedule because they weren't linked, yet in practice these services ended up having a major impact on the progress of the structures.

Since the site was relatively small, with many different structures and services, it became very congested, and the restricted access should have been taken into account at the time of schedule preparation. Trenches excavated for services blocked access to the structures on the site, making it impossible to position cranes, or place construction materials close to where they were required, which all impacted on, and delayed the structures.

It was, therefore, impossible to excavate all the trenches for the services simultaneously since this would have blocked off the whole project site, preventing access to the various structures. The services also had to be scheduled so that the deepest services were installed first. In some cases, different services were laid in a common trench, but at different depths, and should have been scheduled accordingly. Some services were deeper than the adjacent structure and the service had to be installed before it could be constructed. While some services connected to a structure which had to be constructed before the service could be installed, while others went under the structure and needed to be installed first.

It should have been obvious that the services could not be scheduled in isolation of each other, or in isolation of the structures, and that even the best scheduling software was not going to show this interface. You can't just depend on a computer generated bar chart to show how a structure should be constructed.

In some circumstances, the best way to understand the interfacing of the various services and structures is to simply sketch rough cross-sectional drawings with pen and paper or on a white board, and I did this many times to prove to both my manager and the managing contractor that the approved contract schedule wasn't feasible. Further sketches were done with the Planner to work out a new schedule, taking into account the interface of all services and structures. Similar sketches were then used to show the Supervisors the order that the structures and services should be constructed, and to demonstrate why one activity had to happen before the next activity.

Another issue with the power station schedule was that the structures were scheduled independently from each other. The only logic used in the schedule being that the structures depended on teams completing a structure, and who then moved onto the next structure. The schedule did not take into account that the site was congested and most of the structures required excavations, or that it was impossible to excavate all the structures at the same time, because it would have prevented access to other areas. In addition, the deepest structures had to be completed before adjacent shallower structures could be started, since the footprints of many excavations overlapped each other. In fact, in some cases, structures had to be partially built to allow the adjacent structure to start, but, couldn't be finished before the adjacent structure was completed.

The whole power station was underlain by rock which varied from just below finished ground level to two metres below ground level. Most services and structures had to be excavated in rock, which took much longer than it would have in normal ground, which slowed and delayed all areas of work. But since the rock was not known about at tender stage, and its excavation was not allowed for in the contract schedule, it gave us grounds to apply to the client for an extension of time variation.

When a structure was delayed by the additional time taken to excavate the rock it impacted directly on the construction of an adjacent structure. However, because the schedule wasn't linked correctly this delay did not translate into a delay on the next structure on the schedule. In fact, the delay in the services trenches caused by the rock had no effect on the critical path on our contract schedule, even though there was a major impact on the physical works.

In addition to the rock, the managing contractor changed details of some structures resulting in them becoming deeper. This change had a knock on effect on the adjacent structures, resulting in the sequence of structures having to be changed to allow for the deeper ones to be completed first. Without the correct links it was difficult to quantify the impact these changes had on the schedule or the overall duration. The managing contractor viewed these changes as minor, and it was impossible to demonstrate their full impact on the overall completion date using the approved contract schedule.

In summary, when preparing a schedule you have to consider, not only the schedule for the individual structures, but also the impact adjacent structures have on each other. The impact not only relates to the spatial separation and interaction between the various structures and services, but also to their access requirements and how these requirements can impede the works on the adjacent structures, or how the work on one structure can prevent access to an adjacent one thus hindering their construction.

Access can often be a significant problem on many sites, and in many cases it's not taken into account, by either the contractor or the client, when scheduling the works. I've been on sites during the mechanical phase where there have been over fifty mobile cranes on a site that's only a couple of acres in size. These cranes were performing different tasks, for many different contractors, and resulted in numerous clashes and major interface problems.

To complicate the schedule further the managing contractor had stipulated contract milestones for some structures. They had the turbines arriving very early in the construction process, so all effort was focussed on meeting these milestones, and placing them in position.

The milestones for the turbines were actually met despite the rock and the changes in the drawings. However the placing of the turbines achieved little because there was still a substantial amount of civil work that had to be completed around the gas turbines, and the mechanical contractor could not install any of the associated mechanical equipment until this was completed. To be honest the placing of the turbines was disruptive to the civil works, some work had to be halted, and certain excavations had to be filled to allow access for their installation and re-excavated once the turbines were placed. In addition work around the turbines was now hampered because the turbines restricted access, and there was a risk that our machines and equipment might accidentally damage an expensive, critical component.

However, when our schedule was prepared it was assumed the civil portion of the project would be constructed in isolation, and didn't allow for any interface with the mechanical contractor, which caused us further delays. This was a naive assumption since there were clearly early milestone dates for mechanical access, and we should have expected the mechanical works to begin before the civil works were complete. Besides, most contracts stipulate that the contractor will not be the only contractor on site and that they must allow interfacing with other contractors.

The interface with the other contractors can:
- have major consequences on the schedule:
    - there's the direct effect of being unable to access portions of your work while other contractors are in the area
    - the other contractors may restrict access to the work area, and cause delays because the most direct route to the work area is blocked, perhaps making it necessary to use a longer route
    - the route to the work area may be blocked off entirely, temporarily delaying material deliveries to the area
    - it may be impossible to work in an area for certain periods of time while other contractors are lifting materials with cranes
    - there may be other contractors scaffolding in the area restricting work

- increase costs due to delays:
    - when staff and equipment cannot be used efficiently
    - since the work progresses more slowly than anticipated in the schedule, which results in the task requiring a longer duration
    - which may mean additional personnel are required in order to complete the task in the required schedule duration
- (when the other contractor has completed their work), result in it being more difficult to undertake your work (particularly when doing earthworks, because when other contractors install their structures the area becomes more congested, resulting in the requirement to use smaller machines, and do more hand work, to reach areas in and around them, which is less efficient, more costly and requires more time)

It's therefore essential that when the contract schedule is prepared there is a full understanding of the interface with other contractors. Ideally you want to have completed as much work as possible before other contractors start work in the area. But where there are major areas of known interface the schedule must allow for the possible delays and inefficiencies that may occur as a result of this interface. If the interface is unknown you should qualify the schedule specifying what interface and disruptions the schedule has allowed for.

Ultimately the early placing of the turbines achieved no advantage to the overall project schedule. We should instead have approached the managing contractor at tender stage, or even at the start of the contract, and suggested that the turbine installation was delayed until the civil works could be completed in their immediate area, which would have saved several weeks on the overall schedule.

I often see clients and managing contractors stipulate milestone dates for the installation of mechanical items which actually have no bearing on the final completion date for the project, other than to delay works and cause additional costs. Sometimes it would be beneficial for contractors to assist the client by insisting they set realistic milestone dates for the project, which not only takes into account the work, but also the interface needed for follow-on contractors. In many cases delaying these interim milestone dates would often have little impact on the overall project completion dates and even result in cost and safety benefits to the project.

We also discovered that the schedule for the power station was far too detailed, which may sound surprising considering everything I have said above. Yet the schedule was so detailed, with so many activities, that the site staff generally didn't even bother to consult it, preferring to work to their own schedule. It was also difficult to modify when we discovered the problems.

The main lessons from the above case study are that we need to ensure that a schedule is correctly linked. Individual structures should be considered, as well as the effect they have on each other, and we should also ensure we understand the affect that other contractors and their work may have on our schedule.

Introduction

A project schedule should be drawn up for every project, and used to establish the quickest, most effective method for constructing the project. The schedule can take the form of a simple hand-drawn bar chart or be an intricate, detailed schedule, using a proprietary software package, linking and resourcing the activities.

The schedule can be prepared by the Project Manager, or it may be delegated to a member of the team, who could be an experienced Planner. It's important however, that the Project Manager controls the process, and ensures that the schedule not only meets the contract milestones, but is also achievable with the available resources, and that the project can be constructed in the safest and most economical manner without sacrificing quality standards.

Requirements of a schedule

The schedule must:
- take into account the types, skills and quantity of the available personnel within the company, as well as subcontractor's personnel and those who may be available in the market place
- consider the availability and supply of material
- consider the type and quantity of equipment available
- allow sufficient lead-time for the procurement, manufacture and delivery of items
- allow for adverse weather on the project (extended windy periods, high rainfall, periods of extreme temperatures and even cyclones)
- cater for specific client demands and restrictions such as; blasting, reduced work hours, access, other contractors, payments, safety, and other events
- clearly identify the different activities
- be in sufficient detail to enable progress on these activities to be monitored
- establish and identify relationships between the different activities and these activities must then be linked (many times I've had schedules where items are linked, but the relationship is not clearly shown, therefore it's not possible to analyse how a delay on one activity impacts on another)
- allow time to commission new services and equipment
- allow sufficient time to connect to the existing services (these tie-ins may be dependent on the client being able to schedule the tie-in)
- have an activity for the completion of punch list items
- show milestone dates clearly (these include both the access and completion dates, and, where they are different to those in the contract document, these should be clearly differentiated)

Each activity should have a duration, which is normally dictated by the quantity of resources available for the task. Many clients have the misconception that the duration can be reduced by the contractor allocating more resources to the task. This is often not the case. Some task areas are too confined and restricted in working space, and it's physically impossible to put more resources into the area. Also more resources can mean that the contractor's schedule isn't 'resource levelled', resulting in their inefficient use, and additional costs.

It should go without saying that the activities must be sequenced correctly, although I've had clients expecting me to construct the second floor of a building before the first floor, which is obviously impossible in most cases. Similarly, I've seen contractor's schedules indicating structures being built, before the services that pass under the structure are installed.

Resource levelling

All schedules should seek to level resources as best as possible. On most projects it's impossible to say you require twenty workers today, ten tomorrow, fifteen the next day, and twenty the day after that. If the project has twenty today what happens to the ten spare workers tomorrow and the five spare workers the next day? In most cases workers cannot be sent home without pay, so the project has to pay their wages whether they are used or not. On most projects it's also difficult to have, say, twenty skilled workers arrive on the first day and then, on the last day send all twenty workers away. The ideal is to build up the number of each type of tradesman in a consistent, even manner over a number of days or weeks, reaching the peak requirement, then maintain this peak for several days or weeks until the task is almost complete, and then reduce the number to a point when that trade is no longer required on the project.

If the activities on the schedule are resourced it's possible with most scheduling packages to print out histograms for each requirement, whether it's different types of tradesmen or items of equipment. By studying these resource histograms it may be possible to alter the sequencing and durations of the activities in the schedule, to increase the quantity of resources where the histogram shows a low requirement for the resource, and to decrease the quantity of resources where the histogram shows a peak demand.

It will never be possible to smooth the resources out completely to negate all peaks and troughs, but the smoother the histograms are, the more likely it will be that the resources on the project will be used as efficiently as possible.

Long-lead items

The contract schedule must take cognisance of the delivery time for all the components incorporated into the project, as well as the delivery time and availability of the equipment required to construct the project.

Some projects require specialised formwork or equipment, like heavy-lift cranes, dredgers, large items of earthmoving plant, and tunnel-boring machines. Planning must take their procurement, transport and assembly times into account. I would normally schedule these as separate activities, carefully monitoring the delivery progress to ensure the project is not delayed.

Long-lead items incorporated into the works should be included in the schedule. Often each item will require separate activities, such as preparation of the design, design approval, preparation of shop drawings, drawing approval, fabrication, inspections, transport and installation. The client must be aware of the time constraints they have to approve the design and the shop drawings so the approval process does not delay the schedule.

Relationship between time and cost

Project Managers must understand the relationship between time and cost. It would seem obvious that the shorter the project duration the lower its cost. This, however, isn't always the case. Sometimes a project with a short duration results in a congested project site and inefficient utilisation of the project resources, or excessive peaks and troughs in the resource histograms. It's therefore important the schedule is optimised to be as short as possible without compromising the quality and safety on the project, yet also ensuring the project resources are utilised efficiently.

Awareness of client and other imposed restrictions

When the schedule is prepared it must take into consideration all the client's imposed restrictions recognised and accepted in the tender documents. These may include restricted work hours, limited access, working in close proximity to other contractors, or not impeding access and use of the project site to other contractors or the client. These restrictions may impact the duration of the activities as well as their sequencing, often having a major impact on the way a project is executed.

Obviously any client restrictions which were not known at tender stage will be grounds for a variation and an extension of time claim should they impact the duration or sequence of activities.

When the project will be undertaken in periods of rain, severe cold or other likely extreme weather patterns, consideration must be given to scheduling work that may be affected by these weather events to more favourable periods. For instance, major excavations and earthworks activities should be performed in the dry season where possible. While activities like roofing or those that require the lifting of materials by crane should preferably be done out of the windy season. The schedule should aim to close up buildings and make them as weather-tight as possible before the rainy season or extreme winter temperatures set in. Where it's not possible it will be necessary to allow for possible interruptions and delays caused by the weather.

The critical path

The critical path is the most important part of the schedule, so when the schedule's being prepared the Project Manager should always try and analyse the critical path to seek ways in which the overall project duration can be shortened. It's possible to move the critical path so that it passes through different activities by changing the sequencing of the various activities and altering links within the schedule.

Case study:

Another project I was involved with was the civil works for the construction of a coal processing facility. The major component of which was the coal washing building, consisting of about fifty concrete footings which were foundations for steel columns, and several foundations for various items of equipment. The steelwork and equipment installation was not part of our work. The whole building was about forty metres by fifty metres in area with the footings founded just over a meter below ground level. One part of the building, covering an area of about fifteen by ten metres, had twelve concrete columns, six metres high, supporting an elevated concrete slab. The whole building then had a concrete floor slab at ground level graded to falls to various pits. Quite simple really! However the Project Manager did not plan where to start the construction.

There should basically have been only two issues with the construction of this building. Firstly, the areas that required the most work were the columns and the elevated slab at six metres. The critical path of the schedule should have gone through this structure and it should have been started first. Secondly, as the building covered a large area the schedule should have started from either the centre and worked outwards, or from one end and moved across to the other. Yet what I found was that because the columns were the more difficult work these and the elevated slab were left to last, causing the overall schedule for this building to be about four weeks longer than it should have been. Also, because there was no logical sequence to construct the rest of the building, the floor slab had started several weeks later than it could have. But if the work had been sequenced properly we could have ensured that once the foundations were completed in an area we could have started the floor slab immediately. A poor schedule and poor planning not only resulted in the overall construction period being longer than it should have been, but it also resulted in the inefficient use of personnel and equipment.

The important lesson from here is that work on the critical path items should begin as soon as possible.

The Project Manager may also decide to move the critical path from one set of activities to another for strategic reasons, such as their view of the client's capabilities to provide access or information on time, or the contractor's capabilities or the availability of resources required for these activities. Moving the critical path could also result in a better utilisation of personnel and equipment. It may be necessary to rework the schedule a number of times to ensure the best outcome for the contractor and project.

Agreeing the contract schedule

The contract schedule must be prepared and submitted as soon as possible after the contract is awarded, and certainly within the period specified in the contract documents. The Project Manager must follow up and ensure that written approval of the schedule is received from the client (many times the client or their representative only gives their verbal approval of the schedule which is not acceptable).

Without an agreed contract schedule it's difficult for the contractor to submit extension of time claims for delays caused by the client, and it's also problematic to proceed with the work, because the final approved schedule may be substantially different to the one first submitted for approval.

Sometimes clients and managing contractors avoid agreeing contract schedules when they know that they are unable to issue drawings or information on time, or provide access in accordance with the schedule.

Information required schedule

To ensure that the project is not delayed by the client issuing drawings and information late, or by not granting access or permits according to the schedule, it's important that the client is made aware of when these items are required. Very often this takes the form of an information required schedule, which can be automatically generated by most scheduling packages.

To generate an information required schedule the contractor needs to allocate lead-times to the items of equipment and material required. The contractor may only need a week for some items of information (for example to perform simple excavations), while some items that require designing and fabrication could take ten or more weeks to procure, consequently the schedule needs to reflect this.

When tendering for the project the contractor should take into account the lead-times and specify in their submission what their requirements are. Clients often specify lead-times which are too short and if the contractor has accepted these inadequate lead-times it will cause problems with the contract schedule.

The information required schedule should be issued to the client with the contract schedule. It must be updated at the same time as the schedule, and should be done by a responsible staff member who is aware of the type and quality of information required. Often clients will issue drawings in compliance with the schedule yet they are of little value to the contractor because key areas on them are incomplete, 'on hold', or include insufficient information to begin construction. If this is the case the client must be notified.

When presenting the information required schedule with the schedule update, any delays in the receipt of the information should be highlighted, as should the information required within the following two weeks. The contractor should be proactive in regard to obtaining drawings promptly so as to ensure construction works are not delayed.

Where possible these updated information required schedules should be discussed at the project progress meetings.

S-curves and histograms

If the schedule has a quantity measure for each task or activity (for example areas of formwork, volumes of concrete, tonnages of steel, lengths of pipes and so on), it's possible with most electronic planning packages to draw s-curves and histograms for each of these tasks or products. These curves plot the quantities on a time scale making it possible to see what quantities should be produced each day or week to remain on schedule. S-curves and histograms can be useful tools to monitor and assess progress, and when they are distributed to Supervisors it enables them to understand what their teams are expected to produce daily.

These histograms have a number of uses:
- Most importantly they are an indication of the resources required on the project, and the Project Manager can use them to procure the correct types and quantities of resources.
- To smooth out the resource requirements during the preparation of the schedule, and to ensure there are no excessive peaks and troughs in the resources, which could cause poor utilisation with resultant additional costs.
- If the project is behind, but has resources equal to, or in excess, of the resource histograms, then this could be an indication these resources are not achieving the desired productivity, and that the project is possibly losing money.

Monitoring progress against the schedule

Many Project Managers leave the monitoring of the progress to the Planner or another member of staff and then don't check the accuracy of the update. Progress must be assessed correctly. I often see schedule updates showing activities complete which aren't complete, or items incomplete which are complete, or percentages complete not measured correctly. (When assessing the percentage complete I find it useful to refer to the Supervisor doing the work and ask them how many days work they still require to complete the task.)

Planners often update the schedule and show a 'percentage complete' versus 'percentage that should be complete' and this is all that's reported on. In fact Planners often print out the updated schedule without including the columns for the 'base-line start' and the 'base-line finish' of each activity, making it difficult to assess the progress against the approved contract schedule.

The Project Manager should ensure that when assessing progress it's assessed against the critical path. Just looking at the overall percentage completion of the schedule can be misleading. Often contractors complete work, not on the critical path, ahead of schedule, which pushes up the percentage of work complete, but seldom has much effect on the project's final completion date. The critical path is called this because if a task on it is not completed on time, it influences the final completion date, and is usually difficult to recover from.

The best way I've found to monitor progress, apart from doing the update electronically, is to take the contract schedule each week and do a manual update the old fashioned way. That is, on the date of the update, take a pen and a straight edge, and draw a line down a paper copy of the base-line schedule indicating the actual progress of that task. If the activity is behind schedule the line will indent to the left to the point that equates to the actual progress of that activity bar. If the activity has not started then the line you're drawing should indent to the start of the activity bar. If the activity is say 50% complete, the line would be drawn through the midpoint of the bar for that activity, and if it's complete the line would return to the date on the schedule when the update was done. Each time an update is done draw in a new line on the same page but in a different coloured pen.

By doing the update manually, and showing each update on the same schedule print-out, it's easy to see the exact status of each activity. If this schedule is pinned up on the walls of the meeting room or in the Project Manager's office everyone can see immediately where items have not progressed from the previous week. Of course, with a major project, with numerous sections, and a schedule of several thousand activities, this manual method will become too cumbersome. In that case consider getting the different section managers to monitor their individual areas in this way.

On many occasions I've seen the contract schedule updated weekly, yet nobody in the project team has actually interpreted the data. The Project Manager must understand the schedule, where slippage is occurring, its cause, and then take appropriate action as early as possible to prevent further slippage and to recover the lost time. However, when faced with the schedule update as reams of pages, it's often difficult to do this. The Project Manager must therefore ensure that the monitoring of the schedule, and reporting of project progress, is kept as clear and simple as possible, so problems are visible and easily tracked.

Communicating the schedule and progress to the project staff

It should go without saying that the schedule must be communicated to the project staff. However, on many projects the Project Manager and the Planner don't inform the staff doing the work, resulting in Supervisors not understanding what critical tasks must be completed, or the dates when they must be finished. This obviously defeats the purpose of any project schedule.

If the schedule is communicated to the Supervisors it often takes the form of handing the complete contract schedule to them. However this print-out is often bulky and cumbersome, so staff rarely read through all the pages to find the section relevant to them. The Project Manager would do better by communicating only the relevant section of the schedule to the staff responsible for that area.

Many Supervisors aren't interested in a schedule dealing with the details for the next several weeks or months. It's therefore wise to break the schedule down into 'bite-size chunks' that deal only with the work in their section that must happen during the next two or three weeks.

Often it's useful to convert the schedule for each section of works into a pictorial form. For instance, provide a coloured and marked-up drawing to the Supervisors, clearly showing the various sections of the work with the completion dates for the components of that section. This drawing can then be displayed in each of the Supervisor's offices and be used to communicate the targets to workers.

As important as communicating the schedule is to report the progress to the staff. Staff on many projects, work in the dark, with little feedback on whether the schedule is being achieved or not. To overcome this, the same pictures described above, can be used to show how many days ahead or behind the task is.

Float

Float is the amount of time that an item can be delayed before it impacts on the completion date of that section of works or the project as a whole. Float can be an emotive subject as the client normally believes the float belongs to them and any delays they caused that reduce this float (but do not impact the end date of the project) aren't grounds for an extension of time variation. The contractor, on the other hand, believes the float belongs to them, and any delay by the client which impacts and reduces it is reason for a variation.

Often the client makes it clear in the contract documentation that they consider all float to belong to them. As a contractor I would always try and ensure that where possible the float in the schedule is hidden and any visible float is kept to the minimum.

Where possible at tender stage the contractor should also try to ensure that it is clearly stated the float belongs to them. However where there is no clear statement as to who owns it, the norm is that the float is used by whichever party requires it first. In other words, whoever causes the delay first will use the float to cover that delay. Only once the entire available float has been used will a delay become a delay claim.

Variations

All extension of time variations should be presented using the approved contract schedule. If the activities on the schedule are correctly linked and sequenced, and have the correct durations, demonstrating a case for a variation of time can be simple. If the schedule is correctly resourced, and the client has access to this data it should also be simple to demonstrate a case for acceleration and the mobilising of additional resources should these be required.

The variations may be due to the contractor receiving drawings, information or access late, weather events or unforeseen interruptions on the project, or the project scope changing as a result of additional activities or an increase in quantities.

All of these delays must be individually entered onto the schedule which should then automatically show the effect of the delay on the project duration. Some delays may partly run concurrently, these may have to be entered onto the schedule together so that any concurrency is taken into consideration.

Most contracts and clients will only accept delays to the critical path and those that affect the overall contract duration. It's important that the effect a delay has on the critical path is clearly demonstrated. Also as discussed previously, most clients will want to see the schedule float consumed before they will accept that there has been a project delay.

Many clients will also want the contractor to accelerate the works to reduce the impact of the delay on the project completion date. But before the Project Manager considers this, they should first ensure the client accepts the impact of the delay on the contract schedule. Once the quantum of this has been agreed the Project Manager can then prepare an acceleration schedule.

(Refer to Chapter 12 for a further discussion of acceleration).

Approval of a revised contract schedule

Sometimes it's necessary to revise the contract schedule because there has been a change in the project scope (additional work has been added or changes made), delays or variations may have been approved, the client may have changed the order of the works or requested an acceleration, or drawings have been issued for sections that previously had insufficient information to prepare a detailed schedule. Whatever the reason, it's important the revised schedule is approved in writing by the client so it can become the new contract schedule. I have on many occasions experienced the contractor working to an 'unofficial contract schedule' which is obviously meaningless; project staff should work to only one schedule – the official approved contract schedule.

Special use schedules

As well as the contract schedule there may be a need for more detailed schedules to cover a section of the works, a particular trade, sometimes even a particular activity, or the interface between different activities or different contractors. These schedules are not normally contractual, but, in most cases their overall duration should tie into the contract schedule.

Each subcontractor should prepare a detailed schedule for their portion of works which must fit the time allocated for their work on the contract schedule.

It's often useful, near the end of the project, to prepare detailed commissioning and finishing schedules, plus ones to show the completion of the punch list items. These schedules are not necessarily contract schedules, but are used so staff are aware of the order that items should be attended to and completed. Often these are simple hand-drawn schedules prepared by the Engineer or Supervisor responsible for a particular section of works.

Summary

Project schedules are a valuable tool used to plan and sequence the work on a project so that the tasks can be performed in the shortest possible time using the available resources as efficiently as possible.

The contract schedule:

- not only protects the client but also protects the contractor's rights
- should take into account any imposed restrictions, as well as possible weather events
- should allow for the construction of individual structures, as well as taking into account the interface between the various structures and any potential access problems created when work is performed on the neighbouring structures
- should be sufficiently detailed to enable progress to be monitored and updated
- activities must be correctly sequenced and linked with sufficient time allocated to each activity to enable it to be completed using the available resources in a safe manner and to the required quality standards
- must be approved in writing by the client
- should, together with the key dates, be conveyed to the construction staff in a form that can be readily understood and used
- should be updated on a regular basis, with particular emphasis being placed on monitoring the progress of the activities on the critical path
- if resourced correctly, will ensure the required resources will be available on the project
- resources, when compared with the actual resources used, will give an indication of the productivity on the project, and enable the Project Manager to anticipate possible financial problems
- should be used to develop an information required schedule
- if properly constructed, is an important tool that can be used to assess the impact of delays or scope variations

Chapter 4 - Delivering the Project

After thorough planning, and ensuring the project is started correctly, it's essential the project is managed effectively and efficiently.

Delegation

On large, complex projects the Project Manager may have several staff working for them, such as Assistant Site Managers, Section Engineers, Engineers, Safety Advisors, Quality Managers, and Supervisors. It's impossible for the Project Manager to attend to all the tasks, so it's therefore necessary to delegate tasks to members of their team.

I've seen projects where the Project Manager has tried to do everything, often leading to disaster because they take on too many responsibilities, resulting in tasks being completed late, done badly, or forgotten altogether. This leads to frustration in the rest of the project team, since they are left waiting for materials or equipment, which the Project Manager was supposed to organise. Furthermore, members of the team may feel overlooked and not trusted to manage their section of work, resulting in poor morale and productivity.

Frequently I hear excuses as to why a task has not been delegated such as; 'it will take longer to show them than if I did it myself'. Yet how will the other person ever learn if they are not taught? Time spent training the person now, will, in the future, be time well spent, since hopefully the person will be able to do the task unaided.

I also hear comments like the person is 'useless' or 'incapable of doing the task'. Are they really useless, or is it because they've not been shown what to do? Do they have the appropriate training and experience to fulfil the role expected of them? If they are genuinely useless, why are they still employed and what's being done to replace them?

Of course, the opposite of not delegating tasks is to delegate everything, or to delegate inappropriate tasks.

Case study:
On one project, the Project Manager delegated the task of finding subcontractors, and getting them to quote for work, to a junior Engineer. One subcontractor required was for the project roads. However the Engineer had no prior experience of road-works, nor experience in tender and contract documentation, and nobody explained to him how to proceed with the task.

The results of the above were, firstly, the Engineer requested subcontractors to tender who weren't necessarily qualified or capable of carrying out the works, and the Engineer did not define the scope correctly so the tenderers priced different work. In addition, subcontractors were not given all the applicable documents so were not made aware of all the project conditions.

When we received the tenders back they were incomplete because they didn't include for all the requirements. It was also difficult to compare the tenders, since each subcontractor had made their own assumptions regarding what the work *involved. It took several further weeks of discussions with the subcontractors to ensure they had priced everything.*

Who was blamed for this? The Engineer of course! But the Engineer knew no better because the task was not properly explained to him.

If we hadn't discovered the flaws with the tender procedures we would have awarded the contract to the subcontractor who submitted the lowest priced tender, who may not have been capable of carrying out the works, and who would've put in numerous variations once work started and the actual scope of the work became apparent.

When delegating a task ensure the person you are delegating the task to has the appropriate knowledge and experience to carry out the task, and if they don't it's necessary to coach them.

Always check that the person delegated a task understands the requirements. Often a delegated task ends in failure because the person didn't understand what they were supposed to do. Who is the idiot in this case, the person who misunderstood the instructions or the person that didn't give clear and concise instructions?

Also, ensure the person given the work is suitable and appropriate. Often young and inexperienced staff are delegated to deal with a subcontractor, or the client, leading to them being bullied. Sometimes the subcontractor or client may even object to dealing with these people because they are inexperienced, and don't have the suitable levels of authority.

It's important to note here, that even when a task is delegated, you should follow up and ensure the task has been done. To delegate responsibility doesn't mean you can abdicate the responsibility. I've often requested a Project Manager to provide me with information, or carry out a task, and when I later asked for an update the reply is, 'I asked the Engineer to do it, I will call them and ask if it's been done.' I find this unacceptable, since it means the Project Manager has not followed up and checked the task was completed.

Time management

Poor time management is a problem for many of us, including the majority of Project Managers I've worked with. Project Managers are bombarded hourly by problems and competing demands, and if they don't manage their time effectively it will be impossible to deal with all the issues efficiently and timeously.

To help things run smoothly ensure there's an orderly filing system, and that paperwork and emails are filed in the correct place. Importantly, return documents to their correct place after you've finished referring to them. Searching for information is time consuming, and many tasks are left incomplete when information cannot be found.

A notebook is a useful tool to assists with time management. Whilst walking around the project, talking to people, or receiving a phone call, make notes of what must be actioned. By noting all the tasks that need to be accomplished in the day, and crossing them off as they are completed, it's possible to ensure items are not forgotten, are prioritised enabling the most urgent to be dealt with first, and that you to have a clear head, uncluttered by the mass of tasks requiring attention.

It's also important to protect your time. A Project Manager always has multiple issues to deal with, people in and out of the office, and the telephone ringing. Somehow your time must be divided up to attend to everyone's demands while actioning important tasks. Sometimes when the issue or call is relatively minor you may have to be rude (in a nice way of course), and ask a person to contact you again later. Consideration may even be given to working after-hours when the chance of interruption is less. Alternatively, set aside a regular hour every day to attend to specific tasks, and let staff know you don't want to be interrupted during this period, other than for an emergency.

Teamwork

Teamwork is an essential requirement to run a successful project, since no construction project is done by one person alone. There are a number of different teams working together on any construction project, and the personnel within these teams must work in a coordinated fashion, to guarantee the success of the project. To ensure they function correctly it's essential that the Project Manager manages and coordinate the teams.

On a project most personnel work in teams, which may be task orientated, such as concrete placing, formwork erecting, reinforcing fixing, and block laying, and they may also be arranged so they report to a Supervisor or Foreman. These teams must be structured to include the required skills for undertaking the particular tasks, and the individual team members should be able to work together in a harmonious fashion to maximise the team's output. In most cases the Project Manager won't be directly responsible for these teams, however, they should be aware when there are problems within them, like one team member not pulling their weight or causing discontent with other personnel. Where this happens it will be necessary for the Project Manager to take action to improve cooperation, which may involve discussing the problem with the Supervisor, moving personnel around within the team or between teams, taking disciplinary action, counselling the offending team member, and possibly even replacing the Supervisor (if the problem is a result of the Supervisor). Failure of one team can impact on other teams, reduce productivity, and result in poor safety and quality. All of which, impacts on the schedule and the profitability of the project.

It's the Project Manager's responsibility that all staff work together as a team. On many projects there's dissention between different staff members, resulting in poor morale which impacts on productivity and performance. So it's essential the staff making up the management team on the project have the correct set of skills and aptitudes to work equally and cohesively when attending to all the required management tasks.

A successful project is not only about having a good contractor. It also depends on the client's personnel and their support team (Engineers, Architects and managing contractors) being able to work together with the contractor as one combined team. Project Managers should assist with this process, and work with the client's team to prevent and resolve any possible problems. They should not only be part of the project team, they should add to the stature of the team. However, it should be noted, that this requirement does not mean the Project Manager foregoes the right to defend themselves and their company, and it should not prevent them following the project contractual processes, and lodging claims and variations that the contractor may be entitled to.

Communication

Communication is something that a Project Manager has to do all the time, every day. It's both verbal and written, and will be with workers, project staff, management, Head Office staff, subcontractors, suppliers, the client, the client's team (which may include Engineers and Architects), local authorities and members of the public.

Communication should:
- be civil
- be clear and concise
- be persuasive and forceful enough to ensure instructions are followed
- achieve the best outcomes for the project,
- be effective
- take into account relationships
- take into account the level of understanding the other person has
- not be condescending

Good communication is vital to the success of any project. There are courses and programs which can improve the level of written and oral communication, and consideration should be given to attending one of these.

Communication isn't just about giving and receiving instructions, it's also about keeping the various stake holders informed about developments on the project, irrespective of whether they are project staff, workers, subcontractors, suppliers, client representatives, the client or neighbours. The amount and level of communication will vary according to circumstances and the individual. Communication is often best given verbally, which could be at meetings, directly one-on-one, or via telephone. However it may sometimes be in the form of letters or memos that may be addressed to specific individuals, or may be just generalised memos handed to all affected parties, or displayed on project notice boards.

When issuing letters, emails, memos and verbal communications thought must go into the type of communication. Inappropriate communications can cause irreparable harm to the project, individuals and personal reputations. Often things are said, or written in haste, which are regretted long after the event.

On large projects, communications may be addressed to the client, the client's representative, subcontractors or suppliers, and these may be issued by various staff members. It's always important the correct channels of communications are used, and only authorised staff members communicate with the relevant parties. If more than one member of staff communicates with any of the parties it can cause confusion, and result in the other party receiving mixed and confusing communications, which is not only unprofessional but can lead to problems.

All staff must be aware of the communication protocol, and it's the Project Manager's responsibility to check that all correspondence is carried out in a professional manner. All correspondence of a contractual or financial nature should be reviewed by the Project Manager, and letters should never be signed on behalf of the Project Manager unless they have read the contents and authorised its release.

Emails

Emails can save time on a project, but they can also waste time and lead to embarrassing actions which we would all rather forget. For this reason emails must always be dealt with in an appropriate manner.

Case study

One manager's primary mode of communication was via email, and he often sent an email to me, or others, when we were in our offices, less than fifteen metres from him. Usually these emails required a reply, then further discussion, and in the end we'd each write a couple of emails about a subject which could have been resolved with a two minute conversation. How much easier and simpler it would have been to walk across to our offices, pose the question and then discuss the answer

Think before sending an email. Only send them if you want a record of the discussion, if you don't want to interrupt the person, or if you know they are unavailable. Otherwise things can usually be resolved more quickly and easily by talking in person.

Emails waste people's time when they are copied to numerous people. Emails should only be sent to people who are affected by the email, or who are required to respond to the email. I often see personal emails copied to all staff on a project, and sometimes even to the client.

When replying to emails always consider who you're directing your reply to, in many cases you probably do not want to press the 'reply to all' button. I'm sure we have all done this in the past, leading to embarrassing situations when our reply, which may have contained sensitive information, has been distributed to everyone the original email was sent to. Remember personal emails, or emails criticising an individual should only be circulated to the individual concerned.

Emails should be addressed to the person who is required to action the mail. For example, by sending an email to all the Engineers on a project, requesting an item to be actioned, may result in them all taking time to read the email, but nobody taking action, because no one was directly given the task.

Emails can be both time consuming and disruptive, and I've had Project Managers stop whatever they were doing to read a new email. No email is that important that it has to be read immediately. Doing so while attending to another task interrupts thoughts, causing you to lose focus, often resulting in the first task remaining incomplete. For this reason I suggest sounds, or other forms of communication, advising of a new email are disabled.

It's pointless reading an email then forgetting the content, or action required. Only read emails when you can concentrate on their content, and have the time to respond or action them. After reading an email, action it, reply to the originator, or forward the email to the person that must action the email, and then file it in the appropriate mail folder. I suggest times are allocated in the day, to read, and attend to new emails. Consider adding a footnote to your emails stating that you only read and respond to emails once a day.

We have all written things in emails that we regretted later. The problem with emails is that you read an email, reply to it immediately, push the 'send' button, and in an instant your reply is on the screen of the person you replied to. Often the reply may not have been well thought through or considered, it may be emotive, and in some cases it may be rude and hurtful. Rather than sending a reply to an email immediately, consider creating a 'To Send' folder, and put outgoing emails in this folder first. At the end of the day, by which time you may have had a chance to reconsider your response, the emails in this folder can be sent. Of course normal one line replies, or everyday emails can be answered and dispatched immediately.

Letter writing

I've seen many badly written letters, often addressed to the client. In some cases they even involved variations of several million dollars, yet, the Project Manager hadn't bothered to spend time and effort ensuring the contents of the letter could be understood, and that the facts and figures in the letter were correct. In some cases they even misspelled the client's name!

Letters should:
- have a date
- have a reference number
- be addressed to the correct person (The contract normally specifies who that person is, as well as who should be sent copies. If you're unsure contact the company to find out who the right person is.)
- have a heading, including the project reference name and number (letters to the client should use the reference name and number in the contract document), and a second heading line containing the subject matter
- have an introduction, normally a brief overview of the subject within the letter
- include the body, containing the facts and supporting information (where the supporting information is lengthy or includes numbers, calculations and diagrams, consideration should be given to inserting these as appendices, and including only the summary of the documents in the body of the letter, referring the reader to the relevant appendix or attachment)
- have a conclusion which summarises the facts and indicates the required future course of action
- be confined to one topic, or a few similar topics
- be concise and in simple language
- not be contradictory
- not use emotive language
- be checked for spelling and typographical errors (if you know your grammar is poor request someone to check the letter)
- be arranged in easily readable paragraphs
- be numbered correctly and consistently when it's required
- use consistent text (resist the urge to use text that is in capitals, bold, in colour or in italics to highlight a point)
- use exclamation and question marks sparingly
- quote the correct clauses from the contract document, the specific reference from the tender documents or the applicable drawing numbers

- be double-checked to ensure that all calculations and figures are correct and that they tie up

Poorly written letters are often not treated with the seriousness they deserve, and letters which use incorrect facts and figures not only cause the client to doubt the authenticity of the figures, but may lead to a lower value being paid than was requested.

Problem solving

Problems are a daily occurrence on a construction project and Project Managers must be skilled at solving them. It's always important to discover the root cause of the problem. I've had for instance equipment repeatedly break down on a project, with the same problem. Each time it was repaired only for it to break again. The only way to actually solve the problem, and prevent further breakdowns was to discover the real reason for the breakdown and solve this. Sometimes there may actually be more than one cause of a problem.

Equally, there's often more than one solution to a problem, with some solutions being more cost effective and efficient than others. Project Managers shouldn't always think of the obvious causes and solutions to problems, but rather think laterally, to come up with innovative solutions which could benefit the project.

When in doubt Project Managers shouldn't hesitate to ask for outside opinions or assistance in solving difficult problems.

Decision making

A Project Manager has to constantly make decisions. Many will be minor and of little consequence, however, some will be major with large implications for the project, the company and even people's lives. An important part of making a decision is not to procrastinate, it not only wastes your and other people's time, but the lack of a decision, or a clear path of action, may delay the project.

Decisions need to be carefully thought through so that the best option, seen at that moment, is chosen. In the case of major decisions, it may be necessary to request further time for consideration, or to consult a third party, however, in certain cases there may not be the luxury of extra time and the decision may have to be made almost instantaneously.

I've witnessed spur of the moment decisions, on a subject considered minor by a Project Manager, turn into an expensive mistake for the project. On the other hand, Project Managers who have taken too long to consider and make decisions, have also cost projects a great deal of money.

Negotiation

Project Managers often have to negotiate with the client, suppliers, subcontractors, even sometimes with their staff, and an ability to do so successfully is essential.

The negotiations may be; to obtain approval for a method or product, to agree a time or cost variation, or it could relate to the award of further work or a new project. To successfully negotiate, the Project Manager should have all the facts available, be able to convince the other person of the advantages of their path of action and discount their objections. It may be necessary for there to be some 'give-

and-take' in the process, so that there are no losers, and none of the parties feel they have been overridden or cheated.

There are specific courses available which can teach the art of successful negotiation.

Client relations

Having a good relationship with the client is a balancing act, and the Project Manager must strike it somewhere between being pleasant, helpful and cooperative, while at the same time ensuring the interests of the contractor are protected. Many times I've had Project Managers reluctant to submit extension of time claims or variations because they were fearful of damaging the relationship with their client, or because they believed the client wouldn't penalise them if they finished late.

It should be remembered a contract is a contract, and there are strict obligations in terms of what the contractor has to perform and what the client must provide, and when, to enable the contractor to deliver the works. The client will usually have no hesitation in taking steps in terms of the contract, which may include claiming damages should the contractor not deliver the specified product according to the contracted schedule. They will also only pay the contract value, unless an additional variation has been agreed. At the same time, the contractor is well within the rights of the contract to request a variation if the client has failed to meet their obligations, or if the scope of the work has changed.

Although these variations should always be submitted in writing to the client, it's important, the Project Manager regularly talks to the client, and forewarns them when they are about to miss their obligations.

Manage client expectations

It's important the Project Manager manages the client's expectations in terms of the schedule, quality and budget.

Some clients have unrealistic expectations when it comes to quality (I'm not referring to the quality of the construction here, but rather to the quality of the finishes). They may specify cheaper finishes, or products, to save on their budget, yet when these are installed, they may be disappointed with their appearance and blame the contractor for the poor quality. Another problem is cheaper products are often not durable, so the contractor may be called back after completing the project, to carry out repairs and maintenance. For this reason it's important the client clearly understands the type of finish they will get with the products they have specified, however, this should never be an excuse for poor workmanship.

The Project Manager must also ensure the client has a realistic understanding of the schedule and the availability of access for their follow-on contractors. Clients don't always understand all the processes involved with the construction, so sometimes they must be talked through them.

It's important, not to 'over-promise and under-deliver', but rather 'under-promise and over-deliver'. I often see Project Managers commit to actioning, or completing items, by dates which they could never meet, resulting in the client being disappointed when the date is missed. Those who do this time and again eventually lose all credibility.

Clients can equally have unrealistic expectations regarding the cost of additional work, which leads to them being upset when the project overruns their budget, and may even result in them having insufficient funds to complete the project. The Project Manager should continually update the client on the expected final project cost, possibly even erring on the conservative side, providing a higher budget than expected.

Day-to-day planning

Project Managers must check that work is planned correctly, which entails ensuring the correct materials, equipment and personnel are available for the task, access routes and work areas are available, and the preceding work has been completed to enable the next task to begin. Often Project Managers are so busy solving problems and dealing with day-to-day issues of running the project, that they put little thought into planning the next tasks. Some even appear to lurch from one crisis to the next, due to poor and inadequate planning, resulting in them spending even more time solving problems.

I recommend that Project Managers set aside a portion of each day to assess the tasks that must be completed in the course of that day, and ensuring that steps are in place to complete these tasks. They should then consider what jobs should be done the next day, and ensure that everything is in place to enable these to be done. Furthermore, they should consider what tasks need to be done next week, and check what should be done today and tomorrow, to enable those to be attended to. Hopefully, by doing this each day, everything will be in place so tasks can be successfully completed, thus avoiding delays, and the implementation of last minute emergency measures to get tasks completed.

Meeting notes

Take notes of the date, attendees and pertinent points at all meetings in your diary or note book, and keep these past the project completion. Often I've had someone quote a date and a meeting I had with the individual, claiming they'd told me then about something, or I'd agreed to something. After referring to my notebook I was able to verify the meeting took place, as well as what was said.

Its particular important to make notes of conversations with employees relating to disciplinary or safety issues, since these could become vital later should they be involved in similar incidents again.

Computer systems and networks

Most projects use computers extensively so staff must have computers with sufficient capacity to meet the requirements of their tasks. Computers, systems and networks which are slow, or crash frequently, are not only frustrating but waste time, leading to the project missing deadlines.

Specialised software, such as planning packages, are sometimes required. Many packages can be helpful, some however can be expensive so should be managed to ensure only products required are loaded on the computers. Anti-virus software is essential on all computers.

On larger projects consideration should be given to linking all project staff to a computer network, which may even be linked to the Head Office. These networks can be expensive to establish, but there can be cost savings as well as improved efficiencies. The project should have a central document filing system on the computer network, which should be password encoded, enabling only authorised personnel to access the various files. All data must be regularly backed up.

Printers and scanners must be sized correctly for the project. Cheap printers not only result in increased maintenance costs, but also in inefficiencies when staff are delayed waiting for the printer.

There should be policies and procedures in place governing the use of company computers, and all staff must be aware of these. Unauthorised use of computers wastes time, overloads and slows the network, and leads to computer viruses.

File management

A proper and orderly filing system must be maintained on the project. Separate files should be arranged by subject matter, which may include; safety, industrial relations, quality control, contract documents, tender documents, correspondence with the client or their representative, internal correspondence, orders and requisitions, subcontractor correspondence, meeting minutes, financial and cost controls, schedule, and technical data. Each of these files should have demarcated subsections.

After items are received, and actioned, they should be filed in the correct location. When paper copies of documents are removed from files they should be returned to where they were removed.

A master file should be maintained of all correspondence received, or issued on the project. I would recommend a paper copy is kept also. On major projects, various staff members will be receiving correspondence from the client, subcontractors and suppliers. These should be copied to a central receiving person, from where they can be directed to the appropriate staff member to be actioned, after which they can be correctly filed.

Files that contain sensitive information should be secured in a lockable cabinet.

Drawings

The project must have a system to receive new drawings.
- Drawings should be stamped with their date of receipt.
- New drawings must be entered into the drawing register which should:
    - include the drawing title, drawing and revision number, and the date the drawing was received
    - be compared with the client's drawing register and discrepancies reported to the client
    - be maintained even if the client has their own register, since the client's register normally logs the date when drawings left their office, but the contractor may not receive them until several days later
- Drawings must be distributed to the relevant person.
- Drawings issued to subcontractors, suppliers and even staff should be signed for so there is proof of receipt. (On more than one occasion I've had a subcontractor order materials in accordance with an earlier drawing revision and then claim they were never issued the revised drawing). The drawing issue slips should:

- include the drawing number, title and revision
- have the date of receipt
- have the signature of the recipient
- be filed
- A master copy of all drawings must be kept in the project site office. These drawings should:
  - be filed correctly according to drawing number, and if necessary, in their various sections
  - never be removed from the master file unless they are replaced by a revision that supersedes them
  - not be removed from the site office
  - not be defaced or written on
  - be kept up-to-date and superseded drawings should be marked 'superseded' and removed
  - be available on drawing tables where they can be easily referred to
- A master set of all the superseded drawings should be kept in the site office so that drawing changes can be tracked, if necessary.
- The Project Manager must:
  - be aware of recently issued drawings and acquaint themselves with their content
  - be aware of what drawings and information are outstanding
  - ensure drawings are issued to the correct people
  - ensure Supervisors are using the correct revision
- Supervisors should:
  - have access to a clean, dry table to work off their drawings
  - maintain their drawings in a file so they are not mislaid
  - remove and clearly mark superseded drawings
  - ensure they are working off the latest drawings
  - report any problems or discrepancies with drawings

Late information

As discussed in Chapter 3 the client should be provided with the dates of when information is required so as to enable construction to proceed smoothly. This list should be monitored and updated by a competent person to ensure the information is:
- received on time
- relevant
- complete
- accurate

The client should be reminded several weeks in advance of when the information is due, and the updated list should be presented at client meetings.

It's important the Project Manager is aware of the status of all information, and should inform the client via a formal letter when information is late.

Request for information

It's often necessary to request information or clarity from the client, or their representative. These requests must be done in writing, so there's a record of the question and when it was raised. The client may have a standard document to do this, if not, draw up your own form for the project.

The request should:
- include a date
- have a sequential number
- be clear
- reference a drawing number or specification
- avoid having unrelated or multiple requests on the same document
- be addressed to the particular individual who is best able to resolve the issue, or directed to the client's centralised clearing and distribution system, if one is specified
- go through only one or two staff on the contractor's team, so they are coordinated, and multiple requests submitted to the client for the same item are avoided
- be tracked on a register, which must be updated and presented at the client's progress meetings

Shop drawings

Some items installed in the construction process require shop drawings before they can be fabricated. At times it's a contract requirement for these to be submitted to the client's representative for their approval, sometimes even in a particular format and layout. Ensure these requirements are included in the subcontract tender and contract documents. I've, on occasion, had subcontractors submit drawings with the incorrect format which were rejected by the client resulting in a delay.

Delays with submitting or approving shop drawings may cause delays to the project, so it's important that suitable and appropriate staff:
- monitor the production of shop drawings
- check to ensure the item conforms to the dimensions, specifications and quality standards
- submit the drawings to the client for comment, if required
- ensure the client returns the drawings timeously
- return the drawings with appropriate comments to the fabricator
- monitor the corrections
- check and approve the modified drawings
- ensure there's a formal procedure to track the submittal and receipt of shop drawings
- ensure drawings are submitted to the correct person, using the correct address

Many contractors don't check the shop drawings since they believe it's the client or their Designers responsibility. However the Designer often only goes through a 'rubber stamping process', and expects the contractor to have confirmed the item will conform to the project requirements. The contractor may be liable for mistakes on the shop drawings which lead to the fabricated item not meeting the project requirements.

Case study:

On one of my projects the fabricator of a steel frame submitted a shop drawing of the item for approval. Our Engineer failed to check the drawing and simply gave his approval for the work to proceed. The dimensions of the item on the drawing though were smaller than required, so consequently when the item arrived it was too small, and didn't fit the structure. It had to be refabricated, resulting in additional costs and delay in completion of the project.

A copy of all shop drawings must be kept on site. When the relevant item is received it should be checked and compared with the drawings to verify it has been fabricated correctly.

Contact list

The project should have a contact list, containing the name, position, email address and phone numbers of all staff, Head Office personnel, the client's representatives, suppliers and subcontractors. This list should be continually updated, and made available to all staff.

Daily records and daily reports

Daily records, or reports, are important, yet the Project Manager often leaves them for other staff to prepare and submit. These reports may be referred to in the event of a contractual dispute, therefore they must be accurate and, if possible, signed by the client or their representative. Often it's a project requirement to submit these reports daily, and even if it isn't I would recommend every contractor still submits one.

The daily report should record:
- the date
- weather conditions such as the amount of rain, temperature, wind speed as well as the hours that couldn't be worked due to adverse weather
- the site physical conditions (such as encountering rock)
- resources available including; staff, personnel, equipment, subcontractor's resources and site visitors
- work done
- delays and disruptions
- major items of material received
- potential future delays
- any safety, environmental or industrial relations incidents
- any other relevant information

If the client wants the daily report submitted in their format, which doesn't allow for all of the above, or has insufficient space to record everything, it may be necessary to persuade them to amend their format.

The numbers of people recorded on site, in the diary, may be important when the client is adjudicating any claim for acceleration or delays.

It's important when work is performed on a cost recovery basis (refer to Chapter 11) that the number of personnel recorded in the daily report ties-up with the cost recovery records. If they don't agree, the client may only reimburse the contractor for the lesser number.

Often a contractor experiences a delay, and records it on the daily report, but when the delay continues, they neglect to record its continuation, which can cause a problem later, because the delay has been recorded as if it only affected one day. It's important to note every delay on every day that it affects progress.

Weekly and monthly reports

On some projects the client may require a weekly or a monthly report, and sometimes a report must also be prepared for the contractor's own management.

Case study:

On one of my projects the client required a weekly and a monthly report, and after several months these reports had grown to over 100 pages long, which was a problem since it took time to prepare (time that was required for other tasks), making it often late. In addition, the content of the report was not reviewed by the Project Manager, resulting in incorrect facts, contradictory statements within the report, and some statements conflicting with what the Project Manager was telling the client.

This report appeared unprofessional, took valuable management time to prepare, and was unhelpful when we submitted claims for delays caused by the client.

It's always far better to have a brief report that focusses on important items only, is quick to prepare, and which the Project Manager can review easily.

The report should include:
- the name of the project
- the date
- who prepared it
- a brief overview of the status of the project
- status of the schedule compared with the contract schedule, and any concerns with it, including steps being implemented to maintain the schedule
- safety performance
- financial performance
- quality
- current state of the industrial relations on the project including any concerns
- outstanding information
- approved and possible new variations
- manning and equipment schedules compared to the forecast schedules
- status of major subcontractors and suppliers
- any concerns, and actions required to mitigate them

Client site meetings

Most projects have weekly or biweekly meetings with the client or their representatives. If the client has not scheduled meetings, it's good practice to request them. Minutes of meetings should be circulated to all attendees.

Client meetings are often run to an agenda prepared by the client. The contractor should be free to add agenda items which they believe are relevant to managing the project successfully.

Project Managers often go into meetings ill prepared and without information previously requested, resulting in items remaining on the minutes week-after-week. This reflects poorly on the contractor, and the Project Manager, because these minutes are usually circulated within the client and managing contractor's organisations and are often read by senior people within these organisations. Items which aren't closed out tend to irritate the client and managing contractor, giving the appearance that the contractor is ignoring them, and administrating the contract unprofessionally.

It's good practice to make your own notes during the meeting, which not only ensures you have a record of the discussions, but also a list of items requiring attention.

After the meeting, on returning to the office:
- go through your notes
- inform staff of decisions taken that impacts their work
- start preparing for the following meeting by gathering the information requested

When the minutes are issued read through them, comparing them with your notes, and verifying they are a true reflection of the discussions and that no items were missed. The meetings are an important record for the project and could be used in a contractual dispute.

The minutes are usually taken by one party and I find the minutes often don't reflect the discussions accurately. This may be due to an oversight, or because the party minuting the meeting has an objective interest to report facts important to their company, and not to minute discussions less favourable to them. If you disagree with the minutes, or feel that something has been left out, this should be brought to the attention of the person who minuted the meeting. Preferably all corrections should be recorded at the next meeting, but sometimes clients are reluctant to amend their minutes, in which case the contractor should submit a letter with the corrections. Of course you can't go adding in facts not discussed in the meeting, nor should you be deliberately critical, picking up spelling and grammatical errors that have little bearing on the interpretation of the discussions.

The day before the meeting:
- read the minutes again
- ensure you have answers and information for questions raised in the previous meeting
- make notes of necessary corrections to the minutes
- make note of new items that require discussion at the next meeting such as:
    - outstanding drawings or information
    - client delays
    - correspondence not replied to
    - overdue payments
    - any new or potential problems

Site meetings are often ideal places to resolve issues or bring them to a head.

Ensure the correct information is provided at the meeting, don't guess answers if you're in any doubt, and rather request time to gather the correct information (you may have to commit to a time frame for the answer). Many Project Managers commit to delivery dates or completion dates at a meeting, which may be impossible to achieve. These dates are recorded in the minutes, and when they are not met it reflects poorly on the Project Manager and the contractor.

Case study:

I had one Project Manager who would argue nearly every point in the meeting. Often he was arguing about something he was wrong about, and was either trying to prove he was right or trying to justify the mistake by blaming the client. This became tedious, and the Project Manager was so busy arguing these points, he missed important items raised in the meeting which required attention. In addition, as long as he argued the point, saying he would bring new information to the next meeting to justify his position, the items remained in the meeting minutes.

When the item is embarrassing to your company try and close it out as soon as possible, so it can be removed from the minutes. Keep to the facts and only argue points that are worth arguing about.

Staff meetings

Some projects have too many meetings, and I have known some have an hour long Supervisors meeting every day. What were the workers doing while their Supervisors were off site? How many problems on site remained unresolved? How productive was the project? How many risks were taken in this time, which could have resulted in an accident?

Its good practice for the Project Manager to hold meetings with their staff, however these should:

- only involve the relevant staff (those not invited should be informed why their attendance isn't required so they aren't offended)
- have an agenda
- keep the discussion to items affecting the majority of attendees (if a specific task or matter, has to be discussed with only one individual, meet with them separately, and if an individual brings up a topic not relevant to other staff, ask the person to discuss the item outside the meeting)
- be brief, to the point and restricted to 30 - 40 minutes long
- be set at an appropriate time that will cause the least disruption in the day's activities for the attendees (for instance Supervisors should not be called to a meeting at the start of the working day when they are at their busiest organising their teams, it may be more convenient, to schedule their meetings 30 minutes before their lunch or tea break)
- ensure staff have followed through on actions raised at the previous meeting

Staff meetings are useful to:

- advise personnel of progress and the milestones that must be achieved
- update staff of new influxes of personnel, equipment or subcontractors

- discuss concerns relating to safety, quality or industrial relations
- provide feedback to staff regarding problems or issues that affect most of them
- update staff on changes on the project or within the company
- give positive feedback

Reporting procedures

Many companies and projects require lengthy, and complicated, reporting procedures. These take time to compile and are often not read or acted on by the relevant party the report is addressed to.

Reporting procedures should be clear, concise, and highlight important aspects and concerns on the project, however, many of them could be replaced by having a one-on-one discussion with the person. There's nothing more enlightening, than physically walking through the project, talking to the Supervisors, observing and commenting on the quality and safety aspects of the work, and checking the progress and utilisation of personnel, materials and equipment. I sometimes find that managers demand reports in the misguided view that this will demonstrate how important they are.

Photographs

There is truth in 'a picture is worth a thousand words.' Consequently photographs should be taken regularly on the project, at least weekly, with a camera that records the time and date on the photograph. Photographs can be used to record the progress of the work, and they are also useful to record variation work, such as excavations that are deeper than they should be, the depth of rock in excavations, or work that was already completed when a revised drawing was issued resulting in its demolition.

Photographs showing depths, levels or size may require an object or tape measure to show the scale while in others it will be necessary to include the particular machine, or piece of equipment, working on the variation.

Companies often require photographs of the project to incorporate into advertisements, brochures, presentations or newsletters. These could be to show the project in general, or a particularly interesting or difficult structure. Since these need to be quality pictures you may want to get some expert photographic tips. Suitable photographs should be taken regularly during the course of the project, since near the end of the project work may be covered or obscured by other contractor's work. The photographs should show good quality, and work conforming to safety regulations. I once saw a major contractor featured in a prominent construction publication, which included a number of photographs showing unacceptable and unsafe practices. Presumably this wasn't how the company wanted to be portrayed.

Many contract documents have specific clauses relating to the taking and publishing of photographs, and usually permission must be sought from the client. Even if the contract is silent on the matter, it would be good practice to obtain permission from the client to use the photographs.

It's often good for staff morale to include a photograph of the project staff in the company newsletter.

Photographs can be useful to record safety incidents. These should be taken immediately after the incident occurred, before the site has been disturbed too much, and should include the date and time.

Photographs should be downloaded, sorted, and stored in folders to show progress, safety incidents, variations, condition of equipment, quality, subcontractors, and so on, and be saved in date order. It's frustrating to hunt through hundreds, possibly thousands, of photographs which have not been filed correctly, to find, say, a good photograph of the project to include in a publication. Photographs of a poorer quality, or repeats, shouldn't be deleted but rather stored in a separate junk folder, since you never know when you may be looking for a picture showing a particular aspect of the project.

Diaries

In addition to project daily records or daily diaries I would encourage the Project Manager and other staff members – in particular Supervisors – to keep their own diary. This could reflect:
- problems and delays that occurred
- details and attendees of meetings held with:
  - an employee
  - the client representative
  - a subcontractor
  - other stake holder
- tasks done
- equipment breakdowns
- when equipment arrived or left site
- safety incidents
- workers present
- material deliveries
- unusual occurrences and disruptions

I've had some Supervisors in the past who kept detailed diaries that proved very useful when there were disputes with workers, suppliers or subcontractors, or for compiling variations. Of course, Supervisors can turn the diary against you, and delight in quoting exactly what they asked or told you, and when.

Potential problems with the client's design

Sometimes, possibly having more construction experience than the client's Designers, you may notice problems with the design. It's important to always put these concerns in writing, after all should there be a failure of a structure, or a component, the contractor is usually the first to be blamed. Even when the contractor proves it's a design issue, the client often maintains that, 'as an experienced contractor', you should have noticed the design was faulty and not proceeded with the construction.

Case study:
One of my projects involved a large area of concrete floor slabs that had construction joints sealed with a specified joint sealant. The client's Engineer specified the width and depth of the joints, which I noticed were too wide according

to the sealant manufacturer's recommendations. Although I noted my concerns in a letter to the client, we were instructed to proceed in accordance with the drawing.

Nine months after we completed the project, the client notified us the joint sealant had failed, and instructed us to repair it at our cost. I responded by informing them that the design was not in accordance with the manufacturer's specifications, and I attached a copy of my earlier letter where I raised my concerns, together with the sealant manufacturer's recommendations. This was the last we heard about the joint sealant failure. Without my correspondence querying the design, we would have had to return and carry out the repairs, at great cost to us, and if we'd used the same specified sealant the repairs would've failed again.

Sometimes the problem can be much worse than just redoing joint sealant.

Case study:

Our company was involved in the construction of a multi-floor parking garage consisting of concrete columns and floor slabs. It was almost complete, when it was noticed that most of the concrete floors were severely cracked. On investigation the design was reviewed and found to be flawed - and key reinforcing had been omitted. The parking garage was condemned, demolished, and rebuilt at a huge cost. Fortunately the design was not part of our contract, and our company was paid in full for all the work - including the rebuild. Questions were however asked, whether our Project Manager should have had the experience to pick up the problem with the reinforcing design, and whether he had carried out his responsibilities diligently.

There is certainly no harm in raising a query with the design if you have any concerns about it.

Assisting the client's Designers with solutions

Sometimes it's possible to assist the client's Designers by proposing alternative solutions. However in doing so, make sure that any correspondence or meeting minutes, clearly reflects that this is only a suggestion, and the Designer has the duty to verify the new design with respect to its safety and integrity, and accept the associated risks. Some Designers modify their designs in accordance with the contractor's suggestion, and then believe they have transferred the design risk to the contractor when anything goes wrong.

Of course, where possible, we all want to assist the client and their Designer, its good practice, and generally benefits the project, client, engineer and often the contractor too.

Coordination of services

Some projects have numerous services underground or in ceiling voids. It's essential to coordinate these installations to ensure the deepest, and those with limited flexibility (in terms of level and position), are installed first. Failure to coordinate them will lead to services clashing, causing some to be redone to accommodate others, resulting in time and cost impacts, and even affecting worker morale by creating arguments between subcontractors.

Chapter 4 - Delivering the Project

Labour productivity

On many projects poor labour productivity is a major problem. It results in tasks costing more than was originally budgeted, causes schedule slippage, and means more personnel are required than allowed for, in turn increasing accommodation and transport costs, and even requiring additional supervisory staff.

The obvious sign of poor labour productivity is personnel standing idle on site during work hours. This could be due to a number of reasons for example:
- workers not being supervised properly or the Supervisor being poorly organised
- excessive and lengthy staff meetings resulting in workers being unsupervised for lengthy periods and waiting for further instructions, or simply taking the opportunity to take an extended break
- a lack of access to the work area caused by:
    - other workers or contractors working in the area
    - scaffolding being unavailable
    - access routes being closed
- workers waiting for equipment like cranes
- waiting for material, like concrete deliveries
- too many workers on site
- too many workers of a particular trade and not enough of another trade
- one trade waiting for another to finish
- workers being poorly trained and unable to skilfully carry out the tasks or achieve the required quality
- workers feeling unhappy or unmotivated to work

It's important to ascertain the reasons the workers are idle because the costs can be enormous, and furthermore, once workers are in that mindset every day it often becomes difficult to later break the cycle and raise productivity again. I've moved our labour onto site on projects we've started, only to find the client didn't provide the information for work to commence, resulting in our workers being utilised unproductively for the first few weeks. When the client finally issued the construction drawings, it was difficult to get the workers back to full productivity because they were used to being idle.

One group of idle workers can unduly influence others who are working, which results in them also becoming less productive. There's also some truth in the saying that 'a busy worker is a happy worker'. Personnel who have the opportunity to stand around and chat to colleagues often start to see, and cause, problems where there are none.

A common problem is people who don't work the full hours they are paid for. Personnel often start work five minutes later than they should, and then leave the work site five minutes before the official start of the morning tea break, returning to the work site five minutes after the end of the tea break. The same happens at the lunch break, and again at the end of the day. If there are two breaks in the day, then thirty minutes is lost by each worker, and if, say, there are one hundred workers on a project, then a total of fifty hours is lost daily, and in a month of twenty work days one thousand man-hours are lost. Multiply this by the average pay rate on site and you have a large cost.

What this also tells us is, that if one hundred workers are required on a project (working a ten hour day), then an additional five workers must be employed to make up for this lost time (which results in additional wages, mobilisation, transport, and accommodation costs). Of course, to only lose five minutes on each side of a break, and at the beginning and end of the shift, is probably something that happens on a well-managed construction project, the norm is probably closer to ten minutes, although I've seen it as bad as fifteen minutes.

It's important at the start of the project that the Project Manager sets clear rules as to what the expectations are regarding time keeping, because, like other bad habits, it's very difficult to break halfway through the project. It's also essential that the supervisory staff set a good example in regards to their time keeping.

On multi-storey buildings workers often have to queue to use lifts to access their work area and transport their equipment and material. A shortage of lifting equipment is also a problem which affects the supply of materials to the worksite, as can heavy peak hour traffic. Alternate solutions should be looked at to overcome these problems. Many can be prevented if the project is planned, resourced and scheduled correctly at the start.

Poor safety also creates poor productivity because of:
- accidents which result in:
    - time lost investigating the accident
    - the work area being temporarily closed down to investigate the accident
    - the injured workers not being able to return to their full duties or perform their assigned tasks
    - poor morale which influences productivity
- unsafe sites being closed down until they are made safe
- workers being unable to work at full productivity due to adequate safety measures not being in place

Other factors to consider are the distances from the work areas to the toilet, eating facilities, offices, stores and storage areas. Stacking and laydown areas for materials may also be inconveniently located resulting in material being double handled.

Rework and rectifying poor quality are major causes of poor productivity so it's important to ensure work is done to the required quality standards, and that all materials used comply with the specifications.

It's essential that when work is done at night, or in dark areas, that there's adequate lighting, since poor lighting is a safety hazard, leads to poor quality, and tasks not being performed as quickly as they would be in a well-lit area. Workers may also congregate, standing idle, in dark areas where they cannot be observed by their Supervisors.

Fatigue can be a major contributor to poor productivity, and personnel should not work extended periods of overtime, or lengthy periods without having a rest day. I've found that most manual workers should not work longer than a ten hour shift, and working in excess of this results in a significant drop in productivity.

There's generally a learning curve to performing a task, and workers doing similar tasks all day, every day, become proficient, often completing the task more quickly, which in turn increases productivity. Therefore personnel should not be moved around unnecessarily on a project, not just because they have to learn a new task, but also the act of packing up from one task and moving to another is disruptive and results in lost time.

High turnover of personnel is also disruptive, it leads to poor productivity, because new employees are continually being introduced, and these personnel have to learn the tasks as well as the project's systems and procedures.

Productivity is affected by adverse weather such as:
- extreme temperatures
- high winds which cause dust and restrict the use of cranes
- rain which causes slippery and wet conditions

There's not much that can be done to counteract the effects of these conditions, other than moving workers to sheltered areas, working shorter shifts during these periods, and ensuring workers have adequate wet or cold weather gear. (The cost of this equipment is normally less than the costs of the workers being unable to work.)

Sometimes the reasons for the poor productivity may be due to the client because:
- conditions on the site are different from those anticipated at tender stage
- the design is different or more complex from that tendered on
- their operations may be interfering with the construction work
- ground conditions may be more difficult than expected
- their contractors may be hampering and holding up progress, and access to work areas
- they may not have provided utilities or facilities where they should have, or not provided enough
- they haven't provided information or access in accordance with the schedule
- they have made changes to the design, resulting in completed work having to be knocked down and redone, resulting in lost productivity, and possibly poor morale since workers generally hate redoing completed work

When the poor productivity is due to the client the Project Manager should check if there are grounds for claiming a variation.

It's important that Supervisors are made accountable for their worker's productivity. To do this labour productivity on their section should be monitored, and feedback provided to them. Many Supervisors may however just shrug their shoulders, and say they are doing their best, so it's important that the Project Manager is able to suggest ways to improve productivity. Feedback should also be given to workers if the project is losing money due their poor performance.

Plant and equipment productivity

Many of the reasons for poor productivity of plant and equipment are the same as those causing poor labour productivity, but it can also be due to:

- using inappropriate equipment for the task
- using equipment that's too small or too large
- poor maintenance resulting in numerous breakdowns, in particularly if a key item breaks it may impact on other items which cannot be used because they depend on the item (for example, if an excavator breaks down, the trucks it was loading often cannot be used)
- poor scheduling of refuelling or maintenance
- operators not having adequate operating skills
- absenteeism of key operators
- incorrect set-up of the equipment (for instance longer lift times can be caused by cranes parked in the wrong position or them being rigged incorrectly for the type of lift)
- incorrectly matched equipment (for example, the size of excavator should be matched to the size and number of trucks it's loading)
- the length and quality of the haul roads
- too much equipment working in a congested area
- poor supervision
- items not being reported as broken down
- items not being put off hire

Contractor management visits

There are different levels and types of management visits which include your direct line manager, company senior management, or executives and directors of the company. These visits will be treated in different ways, but since you will normally be forewarned, the one common factor is to ensure you and the work area are prepared, and that staff where necessary, have been informed so they are also prepared. Walk the project and check the safety, quality, housekeeping and organisation are up-to-standard.

Before a visit from your immediate line manager make notes of any questions you may have. Since visits are often rushed, with many distractions, I often found myself forgetting to ask important questions.

Prepare yourself for any questions they may have, such as, actual progress relative to the schedule, cost versus allowable, client relations, problems and performance. With experience of working with a particular line manager you will be familiar with their most likely questions. Also have the answers available to any outstanding questions from their previous visit. By being prepared, it will appear you're well organised, and demonstrate you have an understanding of the project, and additionally the visit will usually be completed more quickly, enabling you to return to your duties. Although the manager is not visiting the project to assess staff performance, indirectly the assessment will happen as they walk around the project.

Organise your day so you're able to spend quality time with the manager, since they usually have limited time, and they won't appreciate frequent interruptions. Some Project Managers believe they'll impress their managers by showing them how busy they are, continually being interrupted and dealing with urgent problems. However, most managers are unimpressed by this, and instead get the impression the Project Manager is disorganised.

Escort the manager around the project site and make notes of tasks, suggestions or questions they have. Not only does this look professional, it ensures all the items discussed are followed through. When asked a question you don't know the answer to, don't guess the answer but rather admit you don't know. Managers normally have lots of experience and will soon realise you don't know what you're talking about.

Introduce the manager to members of staff, and make mention of any specific tasks or items they should be praised for. Praising personnel in front of senior management is greatly appreciated by staff.

When senior management, directors and executives visit the project:
- check they are prepared for the visit and they bring the appropriate personal protective equipment (or see that there's sufficient stock of the correct sizes available for them)
- ensure they have directions to get to the project
- meet them on time
- provide them with a visitor's induction
- give a brief overview of the project, including the scope the company is engaged to do (although generally these visitors won't want too many specifics or details)
- if possible take them on a brief tour of the project showing them the salient points, but bearing in mind most senior visitors will not appreciate lengthy walking tours
- it's good practice to inform the client the visit will take place

Client management visits

These visits should be treated as an opportunity to showcase and advertise the company, and to demonstrate the project is being undertaken in a professional manner. The client's representatives will be indebted to the contractor if their management leaves the project impressed with the safety, progress and quality of the project. However if their management has found fault with the work and criticised them for not managing the contractor correctly, you can be sure they will be unimpressed, which will make things difficult for the contractor later.

Advise staff of the impending visit and check the site is neat, tidy and safe. Ensure extra care is taken with tasks undertaken immediately before and during the visit, so there are no safety or quality problems apparent during the visit. Of course, the project should always meet these standards, and generally it does, but somehow a problem often occurs during one of these visits.

Company inter-departmental visits and audits

Larger construction companies may have departments for safety, quality, human resources and other requirements, which send representatives and auditors to project sites. Often these visits are poorly coordinated, with no prior warning, and the person visiting doesn't meet the Project Manager when they arrive, or see them before they go.

The Project Manager should request these departments visit early in the project to assist with setting the systems up correctly. However, they must be made aware of any project specific requirements so they consider both these as well as the company's requirements. The visits should be arranged so they spend sufficient time with the relevant project staff and Project Manager, to ensure the requirements and systems are implemented correctly.

Before the department representative departs, they should report to the Project Manager to go through any deficiencies detected, so remedial action can be implemented. If necessary a follow-up visit should be arranged to ensure the project has rectified these.

The Project Manager should, in any case, be in regular contact with these departments, and encourage them to send personnel to visit the project, providing assistance as required.

Weather damage

Severe weather such as flooding, strong winds, lightning strikes and hail can cause damage to equipment and structures. It's important this damage is reported to the applicable insurance companies as soon as possible. Then take photographs of the damage, and prepare a report on the cause and extent of the damage.

When preparing insurance claims make sure all of the costs are included, such as the direct repair costs, as well as the costs of supervision, accommodation, transport and so on. Unfortunately most insurance only covers the cost of repairing or replacing the damaged items, and not consequential damages, such as those related to time lost on the project.

Project Managers should read and understand the insurance policies, since they will state what events are covered, insurance excesses and the procedures for lodging a claim.

Making changes

As discussed in Chapter 1 it's important to plan and make the correct decisions at the start of the project. Unfortunately we've all made incorrect choices early in the project which have later adversely affected the project. We haven't staffed the project correctly, picked the wrong equipment, or decided on an unsuitable construction methodology. Usually when we do realise our mistakes we often:

- don't do anything
- do too little
- implement changes too late
- make the wrong changes

Making changes to a project halfway through construction is often costly and disruptive, so:

- carefully consider what's necessary to improve the running of the project
- consider the costs and disruptions of making the change and weigh these up against the benefits
- once it's decided to make changes they should be implemented as quickly as possible insuring minimal disruption
- explain to staff why the changes are necessary and what the benefits are
- it's essential problems are detected early in the project life and that the necessary changes are made as soon as possible

Help

Projects often experience problems and it's expected the Project Manager should solve them. Sometimes however the Project Manager doesn't have the necessary experience or knowledge to solve them, they don't have the back-up from their company, or the problem is simply too big and difficult to resolve. Project Managers either:
- ignore the problem hoping it will go away
- try and hide the problem from their Head Office
- try and solve the problem, often in the wrong way, making it worse

It's important when a potential problem is detected which could jeopardise the completion of the project or affect its profitability, that the Project Manager:
- notifies their manager
- asks for help if necessary
- makes contingency in the project's cost report for the problem's potential impact

Senior management usually don't like dealing with problems, but they dislike surprises and failed projects even more.

Opportunities for further work and marketing

The aim of all Project Managers should be to be awarded additional work on the project. Project Managers and other senior project staff are in regular contact with the client, and should use these opportunities to understand what other work the client may be considering, both on the project and elsewhere, so as to promote their company. For this reason it's also essential they have an understanding of their company's capabilities, as well as its current capacity.

Many of the projects I've been involved with have led to a phase two, and in one case we even went on to construct five phases. In fact most of my projects have ended up with a contract value exceeding the tender value by at least 15%. The additional work is often more profitable than the original project, because you're able to accomplish the extra work using the resources already established for the original project.

Talking to subcontractors and local authorities, and by simply being observant, the Project Manager may become aware of other projects under consideration in the area. If the Project Manager is unable to follow-up on potential projects or leads, they should pass the relevant information to their manager so the company can pursue them.

Of course, the best advertising a Project Manager can do is to ensure they deliver a quality project, on time and without any safety, environmental or industrial relations problems, and that they are a pleasure to do business with.

Summary

To successfully manage a project the Project Manager should:
- delegate appropriate tasks, ensuring the entrusted person understands the requirements and has sufficient knowledge
- ensure there is good teamwork on the project
- manage their time effectively

- communicate effectively with all stakeholders
- write clear and concise letters
- make considered decisions timeously
- solve problems quickly and effectually
- negotiate effectively
- establish sound client relations
- manage the client's expectations
- manage and plan the project
- be prepared for meetings
- review and correct meeting minutes
- make and keep notes of all meetings
- ensure reporting procedures are simple and effective
- walk the site regularly to review safety, quality, progress and productivity
- check weekly and monthly reports are focused, to the point, and accurate
- hold regular staff meetings, which are brief and only deal with issues affecting the majority of the attendees
- ensure frequent photographs are taken and filed
- maximise labour productivity
- optimise plant and equipment utilisation
- advise the client of any potential problems with their design
- assist the client with suggestions to resolve problems
- ensure that the installation of services is coordinated
- be prepared for management visits
- prepare the project for client visits
- communicate with the company's internal departments
- ensure insurance claims are reported and processed
- market the company

Efficient project controls must be in place to manage:
- documentation
- drawings, including shop drawings
- daily diaries, which must be accurate and signed by the client
- requests for information

Chapter 5 - Safety and Environment

I have found many Project Managers think safety is not part of their responsibility, however, this is incorrect – it's the responsibility of everyone working on the project.

I would encourage all Project Managers attend additional safety courses to obtain a better understanding of their legal obligations, the safety acts, codes and standards that the project operates under and to broaden their knowledge of safety.

Case Study:

I was a Contract Director for a project in Tanzania and while visiting the site, a scaffold that concreters were working on collapsed, resulting in a worker falling onto the broken timbers. A piece of wood penetrated the upper part of his inner thigh, which bled profusely. Fortunately the client had a full-time paramedic on the project, who treated the injured worker, before transferring him to the hospital an hour's drive away. There was a doctor at the hospital, but facilities were rudimentary, so he made a list of items (like bandages, disinfectant and antibiotics), required to treat the worker and the paramedic went from shop to shop procuring them.

Since there was so much blood, I was concerned the worker would die, and since we were working in a foreign country, I was also worried about what could happen to us. I had visions of the local police arriving, arresting us, and locking us in the local prison, and, I had no illusions as to what the conditions would be like there. It was certainly a scary thought for all of us.

There was, however, a happy ending. The injury wasn't as severe as it had appeared and the worker returned to work a couple of days later. We were fortunate, since I'm sure if the piece of wood had pierced a major artery, which was probably only centimetres away, the outcome would have been very different.

This accident could have happened in any country, and if the worker had died there was the possibility of the Project Manager and I both being jailed if found responsible for the worker's death. In first world countries prisons may be slightly better than elsewhere, and the justice system may allow for a fair trial, however, the consequences of the Project Manager being found responsible for the death of a worker, or member of public, on their project, is still possibly prison time. Unfortunately very few Project Managers understand this, and it's probably the reason many don't take safety as seriously as they should.

But let's put aside the direct consequences to you for a minute, and consider some of the other consequences of an accident on a project. What about the workers themselves? What about their families? Workers have an expectation, and a right, they'll return home from your project in the same health as they started. Families expect to see their loved ones, and bread-winners, return from work at the end of the day, and you, the Project Manager, have the duty of care to ensure this happens.

Over the years I have seen huge improvements in safety on projects. When I look back to my early years in construction, I shudder to think how we used to do things, and the risks we took with our own lives, and the lives of workers. Safety has changed for the better which means we can all sleep easier at night.

However, construction work is a hazardous business and accidents happen easily. In fact, I have often found accidents happen when we least expect them and even when the simplest tasks are being done.

Case study:

On one project we constructed columns twenty-six metres high, as well as a six metre high concrete wall a hundred metres long, using heavy large shutters, which were moved on a daily basis with mobile cranes. The only injury we had on the project was when someone stepped out of his truck and twisted their ankle!

Obviously we, and our workers, were aware of the safety hazards while working at heights, and with the shutters and cranes, so took suitable precautions, but we let our guard down when it came to the simple everyday tasks.

Safety isn't something that happens naturally. Just consider the recent maintenance tasks you did at home or in your garden, I'm sure if you analyse these, you'll find you didn't use the correct personal protective equipment, or follow proper safety procedures. I know I'm guilty of numerous safety breaches every time I work in the garden! In fact, if you think about the last time you were driving, did you obey all the rules? I am sure there were times when you used your mobile telephone while driving, exceeded the speed-limit, didn't stop at a Stop sign, crossed an intersection after the traffic light turned amber, and even took a chance overtaking another vehicle when there was a possibility of approaching traffic. When walking, do you stop at the traffic light and wait for it to turn in your favour, or are you impatient and cross as soon as you see a gap in the traffic? We've all done these things, so why should we expect to do things any differently at work? If you engage in unsafe behaviour outside the work place, the chances are that most of your workforce does the same thing.

Safety is about changing behaviour and changing mindsets.

Case study:

I have unfortunately experienced three fatalities on my projects over the years. One incident was on a building project. The building had a double volume atrium entrance lobby, where workers were using scaffolding three-metres- high to install the ceiling. The scaffold was on wheels so it could be moved around to reach all areas. The scaffold had an access ladder from the ground, handrails, and complied with all safety requirements. The workers also wore personal protective equipment and used safety harnesses when on the platform.

The first floor of offices ran around three sides of the entrance lobby with handrails along the edge. During the course of the day the scaffold was moved to within a metre of the edge of the first floor office slab, when one of the workers on the platform decided to go and chat to a colleague of his working on the first floor. However, since the platform was almost level with the slab, instead of climbing down the access ladder on the scaffold and using the stairs to the first floor, he took a short cut and climbed through the handrails on the scaffold, stepped across the metre gap onto the edge of the first floor slab, and climbed through the handrails onto the slab.

He returned to the platform via the same route, but, this time in negotiating the handrails and the gap, he slipped and fell three metres to the concrete floor below. He was rushed to hospital, but died a few days later from his injuries.

Fortunately an investigation by the government safety authorities found we had all the required safety measures and procedures in place. Still, this is small consolation when a worker has died, and somewhere a family is left without a son, a husband and a father.

This story shows, despite the best safety precautions, if you don't change people's mindset on the project, they will take chances, which often result in accidents. It also demonstrates how easily accidents can happen, and how even a relatively small fall can result in death.

Safety cannot be left entirely to the safety advisors, nor can it be driven only by the Project Manager. Safety is a team effort that needs support from the whole management team and all the workers. To achieve this, the Project Manager has to lead the team – and lead by example.

Safety must be set-up properly from the start of the project, which should be scheduled and planned in such a way that the work can be done safely – the client's deadlines and schedule should not dictate otherwise.

It's also important to analyse the design, and ensure the structures can be constructed in a safe manner. Even if the client is responsible for the design, it's still the contractor's responsibility to comment on it and propose alternate solutions which could result in the project being constructed in a safer manner.

What standard should be set for the safety on the project?

There are nearly always safety standards legislated in the country where you're working. There may also be special standards that apply to the particular industry (for example oil and gas, or mining), the client could have their own standards and the construction company will have certain safety policies, procedures and expectations for the project. Then of course there are your own standards. So which standards should be applied? The answer is, use the highest standards, and always apply standards that ensure people will not be injured or killed on the project.

Often, when I visited a site and queried why the safety wasn't to the acceptable standard, the Project Manager replied, 'the client was happy, and therefore it was okay.' This is definitely the wrong answer! It's vital, not just to comply with the client's standards, but also with all the standards mentioned above. After all, if an employee is killed or injured, it's probably going to be you, not the client, who could be facing prison time. It's going to be your company's reputation that will be spoiled, and it's going to be your company's insurance paying the medical bills. Certainly no judge will accept the excuse that the client was happy with the safety on site so it's not your fault.

Case study:

One project had been running for several months when I was appointed the Project Director. In my opinion safety was poor, with numerous incidents every month, and no proactive measures were put in place to improve the safety. Generally there was no clear safety leadership from our site management. Barricading of open excavations was not clear or consistent, housekeeping was poor, and there was a general lack of compliance with regulations. However, despite this, the client was happy with our safety, and shortly after I started on the project they rated our company's safety performance as being good (although they scored our performance poorly in other areas).

Under my guidance, with cooperation from site management, and with help from the Head Office safety department, we instituted safety audits and checks, reviewed job hazard assessments, implemented safety training, conducted regular safety walks and generally improved the safety awareness of all personnel on the project. In doing this our safety performance improved, and the number of incidents was reduced to 20% of what they were, despite tripling our manpower on the project.

However, the client brought new safety advisors onto their team, who reviewed the overall safety performance and safety rules on site, and found that they, together with the contractors' safety performances, were unacceptable.

When the client reviewed our company's performance three months after the first audit, they scored us poorly. So what happened? We had made huge improvements, however, they weren't good enough because the client had since changed their safety expectations for the project. I might add they were correct to do this. What had originally been acceptable on the project was now not good enough.

If we had set-up our safety systems correctly at the start of the project, in accordance with the norms within our company and the industry standards, instead of setting our standards to the client's originally expectations, we wouldn't have had a problem.

Planning and preparation

Most safety and environmental incidents can be avoided if tasks are planned properly, and possible problems are anticipated and taken into consideration. Personnel must be aware of the risks, and have the correct training, materials and equipment to eliminate or minimise them.

Safety leadership

The Project Manager should be aware that the personnel are always watching their actions, so it's important they are seen to be safety conscious at all times; always wearing the correct and appropriate personal protective equipment, never breaking safety rules, or walking past safety violations without taking corrective action. When on site, the Project Manager should also observe operations to ensure they are being conducted safely.

Rules

Rules should not take the place of safety training and safety awareness. I've been on projects where the client, or managing contractor, has set so many rules regulating safety, that the rules govern the actions of the workers, rather than their safety awareness. By this I mean, we should always stay concerned about whether our actions are safe or not, and not be dumbed down by the rules, blindly following them. At all times personnel must consider their safety, and the safety of the people around them. Just because a rule says you must walk along a particular route, or that you can work in an area, does not mean that it's safe to do so. Everybody should still be aware of machinery, cranes and people working in the area, which may put them in an unsafe position.

Documentation

It's important all safety documentation is kept up to date, and is regularly reviewed. This documentation includes, amongst others:
- the project safety plan
- safety registers
- incident reports
- job hazard assessments
- risk assessments
- prestart inspections
- toolbox meeting minutes
- safety committee meeting minutes
- signed attendance lists for toolbox meetings and inductions
- signed receipts for personal protective equipment
- safety audit results and remedial actions
- hazardous material data sheets

Ensure all current documentation is kept where it's required and is readily available. For instance, it's pointless if the current job hazard assessments are filed in the office, and not at the job site where the physical work is being done, where they can be reviewed by anyone entering the area or carrying out a task.

Case study:

We had a person killed on one of our projects and the inspector from the government safety authority arrived within hours of the accident. Before even visiting the accident scene he spent an hour in our site office, going through the project safety documentation, which he fortunately could not fault. He then proceeded to the scene of the accident and the first thing he noted was the dead workers personal protective equipment.

In the end the investigation could find no fault with our documentation or procedures, and our management was cleared of any negligence. However with hindsight, better worker safety awareness and training may possibly have prevented the accident.

Not only are the correct records and documentation there to protect your workers, but they are also vital to protect you as the Project Manager should there be an accident on the project.

However just a word of warning here, I've visited projects where the documentation is impeccable, but the actual safety on the project is of a poor standard. Documentation alone will not protect people.

I have also been on projects where the safety documentation, safety plans and risk assessments have been prepared and are kept on file, yet the actual workers are not aware of what's written in them because the safety advisors have written the documents in isolation, without involving the workers physically doing the task. This is obviously a pointless exercise! In fact, I have even been on projects where the

Project Manager has not known where these documents are kept or what is written in them!

There are various standard forms used in the preparation of these documents, and if you're not familiar with some of them, you should ask the company safety department to assist, or even approach an outside safety company to help with the preparation of them.

Safety plans

A project safety plan should:
- be prepared before any work starts
- analyse the various tasks
- identify risks associated with these tasks
- be specific to the project
- take into account the interface with the public, other contractors, and the client's activities
- take into account the client's safety plan
- consider the methods and procedures of construction
- should identify mitigating measures the project must implement to ensure the level of risk is reduced to a minimal chance of the event occurring, or, that the result of the event occurring is minimal, both in cost or injury
- should be easily understood
- be communicated to all personnel
- be readily available so it can be referred to during the course of construction
- be updated as the project evolves

Registers

There are various registers to record:
- accidents
- equipment
- electrical tools
- ladders
- slings
- excavations

It's essential that registers are filed, and easily accessible for review and audit. They must be completed by a competent person who must actually inspect the items, not just complete the register sitting in the office. It must reflect items that have been transferred from the project, been damaged, or repaired.

In most cases, the Project Manager delegates the responsibility of completing the registers to a member of their team, however, I would recommend that the Project Manager routinely checks registers to ensure they are up to date, and contain the correct information.

Tagging of equipment

Equipment should have tags which indicate when the item was checked, to confirm it's safe to use, who inspected it, and when the item is due to be reinspected. Other tags are used to indicate items which are damaged or unsafe to use, and these tags should only be removed by a person who is qualified, and authorised to declare the item has been repaired and is now safe.

Toolbox meetings

Toolbox meetings should:
- be held at least biweekly or weekly and more often if necessary
- include all workers, and if it's not possible to hold a meeting with all the workers at one time, the workers could be split into two or more groups
- be about 20 minutes long (this must be strictly kept to)
- run to an agenda
- be set at the same time every week, preferably at the start of the shift
- have a specific safety topic, relevant to the current tasks
- be minuted
- have an attendance register which everyone signs
- be attended by all staff and especially the Project Manager
- be kept orderly and discussions should centre on the safety of the project

An agenda for a toolbox meeting may include:
- a welcome
- a presentation of the safety topic
- a brief question and answer session on the topic (keep these relevant to the topic as in some circumstances workers could deliberately start veering off the topic to waste time)
- feedback on outstanding issues from the previous meeting
- discussion of other safety concerns on the project
- upcoming changes on the project that could impact safety
- compliments for any worker, or team, who has done something out of the ordinary with regard to safety
- updates of the latest safety statistics
- asking if there are any concerns regarding safety, and suggestions to improve it

On certain occasions the contractor may extend an invitation to members of the client's or managing contractor's teams to attend.

Safety meetings

Projects (other than small projects with few personnel), should have a safety committee, which consists of Safety Representatives (workers who are particularly safety aware and are selected by the personnel), the Project Manager and a Safety Advisor. The Safety Representatives should attend training to enable them to understand their roles and obligations. The committee should meet monthly, to discuss and review safety, and to recommend improvements to the project safety. These meetings should be minuted.

In addition to these meetings, most, if not all project meetings (including regular staff meetings, subcontractor meetings, client progress meetings, and those with the

contractor's senior management), should have a section devoted to safety. At these meetings there should:
- be an update of the current project safety statistics (subcontractor meetings would include the statistics for the particular subcontractor)
- discuss current and future safety concerns
- highlight specific upcoming project safety risks
- close out problems highlighted in previous meetings

Inspections

Inspections should be carried out by suitably experienced and qualified personnel, and in accordance with the applicable safety legislation, the project safety plan, job hazard assessments, and when there has been a change in the working environment. Inspections should ensure that items are in a good, safe working condition, and that they comply with the latest regulations.

Permits must be reviewed regularly to ensure they are still valid and appropriate to the tasks being performed.

Safety audits

Safety audits should not be confused with safety inspections. Safety audits are normally conducted by senior management, the client's safety department, the construction company's safety department, governmental departments or an independent safety auditor. Safety audits are done to ensure compliance with the safety regulations and standards.

External safety audits are used to obtain an independent assessment of the project's safety compliance, and should be carried out every few months.

Safety training

Having properly trained and skilled personnel with the required training and licenses is important in ensuring a task is performed safely. Many accidents are a result of unauthorised use of equipment, operators not being trained the correct way to use equipment, or not licensed to operate it, or personnel not understanding or being aware of the hazards on a project.

Many Supervisors and senior project staff do not understand their duties and obligations with respect to safety, and nor do they understand their liabilities should one of their workers be injured. They should therefore, be given suitable and adequate training on safety procedures, project safety, reporting requirements, and their duties and responsibilities regarding safety.

Regular safety training should take place to ensure personnel understand the safety procedures, reporting of dangers and hazards they observe, the project rules, and their obligations to comply with these. All personnel must be able to complete hazard and risk assessments.

First-aid facilities

All projects must have a first-aid kit which is readily available, in an accessible, clearly marked location, close to the work areas. In an emergency, you don't want to be rushing around the site to locate a key, that will open a door to access the first-aid kit, nor should you have to run several hundred metres to the offices for a first-aid kit. In high-rise buildings it's advisable to have first-aid kits located strategically on various floors as the project moves up, and always ensure these locations are clearly communicated to personnel.

The first-aid kit must have sufficient stock and material to enable basic first-line medical treatment to be administered. It should have a register listing the contents, which should be updated as items are removed. The kit should also be inspected regularly, and missing and expired stock replaced.

In the event of an accident

It's important that systems are in place in the event of an accident, and that staff are aware of them.

Importantly, if there is an accident, personnel must have access to a means of communication (such as a telephone or radio), to contact emergency services, especially when working after-hours or weekends. Often the site covers a large area (particularly on road or railway projects), and there may be small teams working in isolated areas, so it's essential to ensure there's access to a communication system that works. It's worth being aware that often on these projects there are communication dead spots where there's no mobile telephone, or radio reception.

Ensure all personnel have access to the emergency contact numbers. They can be displayed prominently on the outside of all office doors, or another solution is to issue all personnel with a card listing emergency contact details, which they can keep on them.

There should be people in every work area trained to use emergency equipment, which would include qualified first-aiders, and people capable of correctly operating fire extinguishers.

As soon as the accident victim has received first-aid treatment, and emergency services have been notified, advise the appropriate managers within the organisation, as well as within the client's organisation of the accident.

Accident and incident reporting

Ensure personnel are aware that every incident (including any near misses which could have resulted in an injury or damage to property), must be reported to their immediate superior.

An accident report is usually issued to the client and the contractor's management, immediately after the accident – some clients and companies expect these to be issued within an hour of the incident or accident. The person within the organisations expected to receive the report will generally vary depending on the type and severity of the accident.

As a minimum the report should have:
- the date and time of the incident
- who compiled the report
- a brief outline of what happened, with sketches and photographs if possible
- a basic assessment of the injury or property damage
- the treatment administered
- the current location of the injured (is the person admitted to hospital,

under observation in the first-aid room, and so on)
- the names and occupations of the injured if possible
- names of possible witnesses
- weather conditions and visibility at the time of the incident
- any other relevant information, such as what has been done to secure the accident area and to prevent a similar incident occurring

Obviously at this stage there has been very little time to investigate the accident fully or to uncover the cause of the accident.

Accident investigation

The accident investigation is a separate document to the accident report covered previously.

It should be done as soon as possible after the incident, and should be undertaken by a competent person. Who leads the investigation, and who forms part of it, will depend on the client's and the contractor's procedures, as well as on the severity of the accident. Some accidents may require a formal governmental investigation, and the Project Manager should understand who the incident should be reported to, and who will be part of the investigation.

In some instances the unions may want to be part of the investigation, or even to conduct their own investigation. The Project Manager must therefore be aware of, and understand their rights, which will often depend on their agreement with the contractor and legal requirements.

As part of the investigation:
- parties involved in the incident, as well as any witnesses, should be interviewed
- photographic records should be reviewed
- the risk assessments and safety plans in place should be reviewed
- servicing records for the equipment involved should be inspected
- reviews must be undertaken of inspection records for equipment, scaffolding, excavations and anything that may be relevant to the enquiry
- any item of equipment that has failed may need to be sent to a specialist for testing to determine the cause of the failure

The investigation report should include:
- the date of the accident
- who investigated the accident
- who was involved in the incident
- what happened
- where it happened
- how it happened
- why it happened
- what should be done to prevent a similar incident from occurring again
- any disciplinary action that may have been taken as a result of the incident

It's important that proper written records are kept of the investigations, and that they are filed and stored in a safe place after the project is completed. A worker injured on the project may, many years later, develop further medical problems related to this injury which could result in legal action, and the contractor will then need to access the information relating to the accident.

The purpose of the investigation is to determine the root cause of the incident so that corrective action can be taken to prevent a similar incident from occurring again. It's often useful to distribute its findings to other Project Managers within the company so they can implement measures to ensure a similar incident doesn't occur on their projects.

Ensure that both the accident report and the accident investigation report are circulated to the correct parties within the contractor's organisation, as well as the client and managing contractor's organisations.

On completion of the investigation the findings can be discussed with the project workers at the toolbox and safety meetings. Include the outcome of the investigation, and steps implemented to prevent a similar incident from occurring.

Safety communication

Safety communication takes many forms, such as inductions and meetings. The company may also issue personnel with a safety pocket book, or card, listing important safety rules.

All projects should have a safety notice board which is used to communicate the latest project safety statistics, emergency contact details, minutes of the toolbox meeting, minutes of the latest safety meeting, the names and photographs of the Safety Representatives and first-aiders, as well as the company safety policy.

Safety posters covering relevant safety topics should be placed on the notice board, and other prominent areas of the project, such as the common office areas, ablution facilities, and lunch areas. These should be replaced and rotated throughout the project on a regular basis, to ensure that they maintain worker interest and topics remain relevant to the current tasks.

Safety signage

(Refer to Chapter 2)

Safe equipment

Equipment must be maintained in a safe, working condition, it should be checked before use, and any damage reported. Damaged items should be 'tagged out of service' until they have been inspected by a competent person, and passed safe to use.

When not in use, equipment must be stored correctly, to ensure it's not damaged.

Equipment must only be used by trained and competent personnel.

Traffic

The movement of vehicles on the project and access roads should be planned and controlled. This starts with careful arrangement of the roads, laydown areas and delivery schedules. Poorly planned routes can result in vehicles interfacing with pedestrian traffic or travelling through congested work areas.

All drivers must be licensed and competent to operate their vehicles and machinery, while all vehicles and equipment must be roadworthy and checked daily.

All drivers and operators, including for delivery vehicles, must be made aware of the specific project rules and conditions, and they must obey these as well as public road rules. I've often had unroadworthy delivery vehicles arriving on a project, and in some cases being operated by an unlicensed driver.

Some projects may also require delivery vehicles to be escorted within the project boundaries.

Regulatory signs, such as Stop signs, speed limits, cautionary signage, and directional signs, must be clear and comply with regulations. Delivery drivers must be made aware of the routes they must follow, otherwise it could happen that vehicles take a wrong turn on the project, resulting in them entering restricted areas, or becoming stuck.

Housekeeping

An untidy project is invariably an unsafe project. You've probably heard this many times, but it's usually true. Materials, tools and equipment, stored and stacked untidily, results in dangerous situations and often accidents. Mainly because:
- they create trip hazards
- they force workers to take alternative routes to by-pass the obstructions, making these routes congested and dangerous
- poor stacking can result in the materials becoming dislodged and toppling on to people or equipment
- tools and equipment poorly stored, or left lying on the ground, may be damaged by being stood on, driven over, or water and dust entering them, which causes damage and may make them unsafe
- materials and equipment badly stored on scaffolding and elevated structures can fall off, damaging the item, the equipment they fall on, and possibly injuring or killing people below, so they should be stacked neatly, where they cannot be accidently bumped and dislodged from their positions (even an item, such as a bolt, dropped from height can cause serious injury)
- loose debris can become missiles in windy conditions causing damage and injury
- emergency and evacuation routes can become obstructed
- they can create a fire risk

Good housekeeping is not only essential for safety, but often increases efficiency because materials and equipment is readily available, can be easily located, and leaves walkways and work areas accessible and easy to work on. It also creates a good impression with the client and their representatives, and gives the impression that the project is well organised and managed. I have also found when a safety inspector, or client, walks onto a neat site, they are less likely to take a critical look at the safety, yet, should they walk onto an untidy and disorganised site, it's almost certain they will look at the operations with a critical eye, finding even the smallest safety violations!

Enforce good housekeeping from the start of the project because it's difficult to change people's poor habits part way through. It's often just as easy to stack an item neatly, as it is to throw it randomly on the ground, and it doesn't take any longer to put an item of equipment back into its storage container than leave it lying around.

Material deliveries, storage and handling

Planning access ways, laydown areas and storage areas at the start of the project, will assist with maintaining the site safe and tidy. Ensure materials are stored and stacked safely in an easily accessible location where they don't block construction activities, create a hazard, can't be knocked over, and will not be damaged by vehicles.

Investigate how materials will be delivered, since items that are palletised can be more easily offloaded, stored, and moved on site, and although pallets may be an additional cost, there will be benefits of increased productivity and improved safety. However care should be taken when unpacking pallets. Pallets should also not be stacked too high, or in such a way they easily become unbalanced.

Case study:

One of my projects had worked over a million man-hours without a lost time injury, and was nearing the end. The edges of the roads required precast concrete kerbs, which were nearly a metre long and weighed over fifty kilograms each. They were delivered on pallets, stacked five rows high by six rows across, and were secured to the pallet with shrink wrap plastic during transport and handling. The pallets were offloaded close to where they were required. Workers removed the plastic wrapping, took the kerbs off the pallet, and placed them in position. However while the workers were removing the plastic they accidently dislodged one of the kerbs at the top of the stack, it then fell onto a worker's foot, and even though he was wearing the correct protective footwear, it didn't prevent his foot from being broken.

In some cases it may even be advisable to order special containers to safely store tools, equipment and material.

Hazardous materials

Some materials are hazardous and should be stored in special store areas, which may require a bund (to capture spills or leaks), ventilation and security. These hazardous materials must have material data sheets that outlines the nature of the hazard, how the material should be used, what precautions should be taken when handling and working with the material, as well as details on procedures to follow should someone accidently come into contact with the material. A copy of these data sheets should be kept in the safety file in the project office, a copy should also be available at the place where the product is being used, and attached to the risk assessment applicable to the task, so that all workers handling the product are aware of the hazards.

It's important that all containers are stored upright and are clearly and correctly labelled, with the product details and classification.

Many products may be flammable so should be kept in a special well-ventilated store, not exposed to naked flames, or used in confined spaces, and always with firefighting equipment close at hand.

Weather

Changes in the weather conditions can result in both a safety hazard and an environmental problem.

Risks from rain:
- access platforms and work areas become slippery and dangerous to work on
- roads become slippery which may cause drivers to lose control of their vehicles resulting in accidents
- roads surfaces may become soft resulting in vehicles becoming bogged creating a traffic hazard, and in addition the operation of extracting the vehicle results in another unnecessary, dangerous operation
- roads with inadequate drainage result in pools of water which may cause vehicles to become stuck, or aquaplane out of control
- heavy rain obscures visibility
- ground conditions become soft, which can cause scaffold supports and crane outriggers to settle
- flooding can destabilize sides of excavations causing them to collapse
- rain water causes erosion and also results in the deposition of silt

Risks from lightning:
This serious problem is often not considered on projects. Many sites are in isolated locations where the cranes and structures may be the highest items in the surrounding areas, and attract lightning strikes during an electrical storm. Projects in an exposed area with frequent lightning strikes should install a meter to monitor lightning activity, and a procedure should be in place to evacuate staff from the work area when necessary.

Tall structures, such as tower cranes, should be properly earthed so a lightning strike can be safely dispersed to reduce the risk of injury and damage to the equipment.

Risks with wind:
Wind can be a major hazard when lifting objects with a crane, and personnel responsible for these operations should be aware of the safe maximum wind speeds, which may vary depending on the crane and type of load being lifted.

Wind speeds can vary drastically within the project depending on the height above ground, as well as the location. Some structures on the project may cause a wind tunnel affect increasing its effective speed. Other areas can be relatively sheltered yet when a suspended load is lifted out of this shelter it's exposed to the full force of the wind.

Wind force on suspended loads can result in the load being pushed out beyond the safe lifting radius of the crane, resulting in the crane being overturned, buckled, or damaged. The wind could also cause the load to spin uncontrollably, causing damage to property or injury to workers.

It's advisable that all crane booms are fitted with a wind anemometer, so the wind speed is recorded at the highest exposed point. All suspended loads should be secured with tag lines so the loads can be secured and controlled by trained personnel.

Wind can blow people off their feet, off scaffolding, or as occurred on one of my projects, even into excavations.

Wind causes dust which results in poor visibility which can lead to traffic accidents. The dust irritates eyes and may be harmful if breathed in. In extreme cases the dust can cause slippery surface conditions, and also enter and damage moving machinery. To minimise dust, the project area should be wet down regularly, and earth stockpile areas should be covered or revegetated as soon as practical. In windy conditions it may be necessary to reduce the amount of traffic, and curtail operations where earth or aggregates are being moved or dumped. In extreme conditions it may be necessary to close the project down.

In addition, wind can cause damage to both temporary and permanent structures. Permanent structures are designed to withstand forces in their completed form, but the Designer often doesn't consider that the structure can be unstable in the partly constructed form. To protect against this instability the contractor should consider installing temporary braces.

Case study:
I heard of one contractor who was constructing a concrete reservoir 50 metres in diameter with an eight- metre- high perimeter wall, who had constructed about 40 metres of wall when a strong wind blew the completed section down. In this configuration the walls were unstable because their strength depended on the full circle of wall being completed.

Not only was it costly to replace the walls, but it added several weeks to the schedule, and more importantly the falling walls could have resulted in serious injury or death.

I have known wind to blow over stacked material, resulting in injury and damage to property.

Wind can cause serious damage when objects become airborne. Ensure that building material isn't lying loosely, but is properly stored, stacked and secured, particularly at the end of the shift. Roof and wall sheeting should be properly secured before moving onto the next sheet. Often workers initially only install the minimum number of fastenings before moving onto the next sheets, and return later to complete the remaining fixings. If however they are interrupted, or the shift ends, they may forget to install the remaining fastenings, so if a high wind came along it could easily strip the roof sheets, turning them into missiles which could cause serious injury.

Risks of cyclones, typhoons, hurricanes and tornados:
Since these can cause severe damage and injury, projects in areas prone to cyclones, hurricanes, typhoons and tornados should have access to the latest weather data so they have sufficient warning of an approaching storm. All offices, structures, and materials must be able to be secured and tied down in a short period of time, and the buildings should be designed to withstand the effects of these storms. Sufficient shelter, compliant with codes and specifications, must be in place to safely house workers in the event of a storm.

Risks of heat:
This can be a major problem in some climates leading to heat stroke, heat

exhaustion, and even death. Heat stress is not only a result of the climate, but may be caused by doing hot work in confined, closed spaces.

In hot and humid climates, personnel should first be acclimatised to the conditions before undertaking heavy manual work. They should be made aware of the early signs of heat stress so they stop work and seek treatment should they experience these symptoms. Personnel must also have access to sufficient quantities of cool water, and have the opportunity to take more frequent rest breaks.

Risks of cold:

Severe cold can result in frost bite and disorientation. Personnel, working in exposed, windy and cold conditions must not only have access to a sheltered, warm environment where they can take frequent rest breaks but also access to sufficient warm protective clothing.

Fatigue management

Fatigue not only reduces productivity, but also causes accidents and safety incidents. Fatigue can be due to a number of reasons:
- personnel working long shifts, or working on designated rest days due to:
    - the project working longer hours to make up lost time
    - personnel electing to work additional hours to earn more money
    - people working extended hours so they can take time off at a later date, perhaps for an extended weekend
    - insufficient workers on the project, resulting in a high work load,
- personnel taking insufficient rest breaks in the course of their shift, due to:
    - operational reasons (for instance workers may not be able to stop doing a task until it has been completed)
    - personnel electing to forego the designated rest breaks to complete their shift earlier
- working in hot and humid conditions
- performing a repetitive and boring task over an extended period
- doing a physically hard task
- working night shifts which disrupts worker's body clocks and sleep patterns (they may also be unable to sleep properly during the day due to high temperatures, or noise around their accommodation)
- the workforce having excessively long, daily commute times (sometimes workers travel over an hour to and from the project each day, making their working day even longer than it should be)
- workers have a lengthy commute returning to the project after their rest day (they may spend several hours travelling from home, often leaving early, or in some cases travelling through the night)
- personal problems resulting in an individual not sleeping at night
- illness, which causes fatigue, and may also disrupt sleep
- partying late at night, not only resulting in these people being fatigued, but also disturbing other workers

To reduce fatigue, the project should:
- resist the temptation to work longer project hours or additional shifts
- staff the project correctly

- not allow personnel to work extended, or additional shifts, out of choice
- provide more frequent rest breaks during hot and humid conditions
- personnel performing physically demanding, or repetitious, tasks should be rotated to other tasks, work shorter shifts, or take frequent rest breaks
- arrange accommodation as close as possible to the project, and where there is no alternative to a lengthy commute, consider working shorter shifts
- after a rest day the project could work a shorter shift to take into account long commute times from home
- insist workers return to the project accommodation on the last day of their rest period
- management and supervisors must be aware of changes in personal behaviour, or illness, which may affect a person's ability to work
- implement noise restrictions in the accommodation

Fire

Fire can be a serious hazard and there have been cases of buildings under construction burning down. To prevent fires:
- reduce and control the amount of combustible materials in an area by:
    - ensuring waste material is cleared off site as soon as possible
    - keeping the site neat and tidy
    - stacking materials properly clear of work areas
    - storing flammable liquids in suitable containers in special enclosures
- hot work must only be done in safe conditions, by trained personnel
- place fire extinguishers at readily accessible locations, and have them checked, serviced and replaced as required, ensuring staff are trained in their use
- where projects are located in vegetated areas personnel must be aware of the fire risks, keep plant growth away from offices and work areas (be aware vegetation can often grow back quickly), maintain fire breaks and plan escape routes
- smoking should be restricted to designated areas

Injury to a member of the public

The Project Manager is responsible for the safety of their employees, the subcontractor's employees on the project, as well as for members of the public who may be affected by the work.

The most obvious precaution to protect the public is to prevent them from entering the site during and after working hours. Preferably there should be a security fence with lockable gates surrounding the site, and clear signage prohibiting access. If the gates are open, they should be manned by a security person, and all visitors requested to report to an office next to the entry gate, where they should sign in before being escorted to where they are going. Whoever leaves the project last, at the end of the shift, must secure the site. People, and particularly children, are curious, and if there's unrestricted access onto the project someone may wander onto the site without the proper safety clothing, equipment, and unaware of the safety risks and hazards.

No project wants a member of the public to fall into a hole or hurt themselves on the site, because even if the entry was unauthorised, the project, and the Project Manager, will be accountable for the injury, and possibly face prosecution. In addition, the injured person could sue the contractor and Project Manager, for thousands, and even millions of dollars.

Often projects are in congested cities, with restricted working space around them. In these cases it's vital to ensure that passing members of the public will not be injured by any falling objects. Remember an item falling off a scaffold, may hit another part of the scaffold, causing it to ricochet several metres, so consideration should be given not only to protecting people walking directly below the scaffold, but also those several metres away.

When interacting with public vehicles its essential traffic signage is clearly displayed, complies with the road ordinances, and that construction areas are well lit if they are used at night.

Traffic entering or leaving the project can cause a traffic hazard, and it may be necessary to install warning signs, speed control signs, additional turning lanes, warning lights, or traffic controllers at these intersections. Entry and exit points should have clear visibility, and vegetation, material, parked vehicles or new structures should not obscure oncoming traffic.

Safety in operational facilities

Many projects are within operational facilities or mines requiring a number of considerations that must be taken into account, therefore:
- Ensure construction personnel working within the operational area are safe, particularly when they interface with the client's equipment and operations and personnel should:
    o be given a full induction explaining the dangers within the facility
    o be provided with the specialist safety equipment required, for example; fire retardant or chemical resistant protective clothing, specialised breathing apparatus and hearing protection
    o be immediately informed of any accidents or planned changes in the facility's operations (this may include the release of dangerous gases or discharges of chemicals
    o be aware of the client's emergency procedures and evacuation plans
- The client's workers in the facility must not be endangered by the construction activities, therefore:
    o the construction area must be correctly barricaded to prevent accidental entry
    o where facility workers have to enter construction areas to carry out maintenance or operations, ensure there are clearly marked and signposted safe entry and egress points
    o it may be necessary, for the client's personnel who enter the construction area, to undergo a special induction to acquaint them of the construction dangers

- they should know who to contact for authorisation before entering the construction area
- they should sign onto the construction job hazard assessments for the area, making them aware of the current hazards, any changed conditions, and what precautions to take,
- should the conditions change in the construction area this should be conveyed to the appropriate facility Supervisors, changes could include moving access points, walkways and egress points within the construction areas, or when construction Supervisors, equipment or personnel are changed
- if the facility is working at night, ensure construction areas are well illuminated so barricading and demarcated areas are clearly visible
- at the end of the construction shift, all barricading must be closed and inspected, all signage must be in place, and the illumination should be operational (often workers forget to return the barricading to a safe position at the end of their shift, sometimes resulting in a person accidently entering the construction area)

It should be remembered that many facilities are operational at night, on weekends, or after the construction workers have left site, and sometimes the client may unexpectedly enter an area to carry out repairs and maintenance.

Substance abuse

This is covered in Chapter 6, however, it's important to stress that the use of alcohol and drugs can be a safety hazard on the project. Not only because they impair the worker's ability to operate machinery and equipment safely, but because they may also cause them to make poor judgement decisions, resulting in accidents and injury.

There should be a zero tolerance to substance abuse. Management and supervisory staff should be aware of the signs of use, and there should be regular, random, testing done on the project. Posters creating awareness of the dangers surrounding substance abuse should be displayed, and at least one toolbox meeting should be devoted to this topic.

Environmental

Case study:

One of my projects involved the construction of a large electricity substation which covered several acres of hillside in a farming area. Part of the works was to construct a simple cattle fence around the site boundary, which passed through trees and bushes, went down through several small valleys and up the hillside. At the time, we were constructing the earthworks platform for the substation, and had several large pieces of earthmoving machinery on site. Our fencing subcontractor decided the terrain the fence crossed was too rough and overgrown, so to make his life easier, and unbeknown to us, he arranged with the operator of the dozer (which was a D10 – in other words, a large dozer with a blade several metres wide) to work on Sunday and clear the fence line. The fence line was situated exactly on the boundary, and the dozer cleared a strip of ground four metres wide, but since the fence line ran down the middle it was effectively encroaching at least two metres into the neighbouring property. Needless to say, early on Monday morning I had a very unhappy client storm into my office, saying the neighbour had complained of the environmental damage caused to their and the client's property.

Grass, trees and shrubs had been stripped from a four metres wide area two kilometres long, and the debris were pushed down the valley, causing further damage to the surrounding vegetation and neighbouring property. Also because the clearing was done on the side of a hill, an erosion problem was created when it rained.

The clearing was clearly in breach of the client's approved environmental plan, and they faced the risk of being fined, and even having to stop the project.

We had to reassure the adjoining property owner that the clearing was an accident, and that we would take every care to reinstate the area, and ensure something similar didn't occur again. Using smaller equipment we cleaned up the debris, reinstated and opened the drainage lines, and constructed new berms and drains to prevent rain runoff from eroding the stripped areas. Then we reseeded the cut areas with indigenous grasses and watered them until they had grown sufficiently to cover the dozer scars.

The matter was embarrassing to us, but it could have been far more embarrassing. We were fortunate the neighbour didn't take the matter further, and that the environmental authorities or local newspaper didn't find out about the incident. But it did cost us several thousand dollars to rectify, and when I say rectify I should say patch, because with an environmental accident whatever you do is normally just a patch. It's nearly impossible to restore the area to what it was before a dozer flattened it, trees take years to mature, alien vegetation takes the opportunity to establish itself, water finds a new path to travel, and the freshly unearthed rocks take years to weather and match the surrounding rocks.

This story illustrates how easily and quickly an environmental incident can occur, and how much angst and cost can result from it – all because a fencing contractor wanted to make their life a little easier

Lessons to learn from the above are firstly, the boundary fence should not have been constructed exactly on the boundary, which may have prevented damage to the neighbour's property. But more importantly, no one should have been working unsupervised after-hours. All workers and subcontractors should be inducted to the project, which should include a section on environmental aspects, and in particular, those dealing with the site boundaries and areas of environmental concern.

Waste handling and disposal

Systems must be put in place to collect and handle waste generated by the project. Many projects generate a large amount of waste, which includes general rubbish, paper, packaging, building debris, waste concrete, liquids, paints and solvents. Some of this waste can be dangerous, while most creates an environmental hazard.

The costs of disposing of waste on a project can be significant both for collecting, loading and removing and the dumping fees at the local tip. But by separating and recycling this cost can be reduced. It's important though that all project personnel, including subcontractors, are aware of the recycling initiatives and that this is enforced as far as possible. Unfortunately on remote sites it may not be feasible, or possible, to send waste material for recycling.

Other ways to reduce the waste include:
- minimising rework
- handling materials in such a way as to minimise breakages
- eliminating wastage of materials
- ordering the correct quantities of materials
- crushing building waste and waste concrete on site and reusing as earth fill material
- returning packaging to suppliers
- requesting suppliers to minimise packaging
- using alternative products that are more environmentally friendly and easier to clean up and dispose
- using companies that collect and recycle used engine oils

Washing of concrete trucks, concrete pumps and concrete placing equipment can generate large quantities of contaminated waste water, so an area should be set aside for these operations, designed in such a way that water usage is minimised and where possible dirty water is collected and reused. The waste cement must be collected and disposed of in a safe manner.

Fuel, oil and other liquids
- Fuels and liquids should be stored in containers in an area with bunds to contain any spills and leaks.
- When refuelling vehicles, or refilling containers, drip trays should be used to contain any spills.
- Accidental spills must be controlled to prevent further spillage, contained to prevent them spreading, then cleaned up using a spill kit, with the waste disposed of in a suitable controlled waste disposal container.
- All spills should be reported and recorded.
- Oil and fuel leaks must be fixed immediately.
- Waste liquids must be properly disposed of in an environmentally controlled and responsible manner.

Summary
- Project Managers must have an understanding of safety legislation, as well as the safety standards and rules required by the client and the contractor.
- Safety rules and standards must be implemented from the start of the project.
- A safety and environmental plan must be prepared, taking into account the project activities, and address any interface with the public, the client, and other contractors.
- All personnel must be aware of the safety plan and the procedures, and precautions that must be implemented.
- Incidents and accidents on a project can result in lost production, damage to property, injury, and even lead to the prosecution of the Project Managers and the company's senior management for negligence.
- Procedures and construction methods should take into account changes in

the weather.
- Fatigue often leads to accidents and personnel fatigue should be managed.
- The Project Manager must lead safety behaviours, however, it's everyone's responsibility to work safely.
- Substance abuse should be prevented, and measures put in place to prevent personnel under the influence from accessing the project.
- Emergency equipment must be readily available, in working order, close to work areas, and personnel must be trained in its use.
- A means of communication must be available in the event of an emergency.
- Systems and processes must be in place in the event of an accident.
- Incidents and accidents should be investigated to establish the cause, and to enable mitigating measures to be implemented so as to prevent similar incidents from occurring.
- There should be clear signage on the project.
- Safety information should be communicated at toolbox meetings, prestart meetings, on project notice boards, and via posters.
- Safety documentation, and registers, must be kept up to date, and all documentation should be readily accessible, and neatly filed.
- Workers must be properly trained.
- Equipment should be safe, and maintained in good working order.
- The project should be neat and tidy, and rubbish must be removed regularly.
- Materials must be stored and stacked, neatly and safely.
- Care should be taken to protect the surrounding environment.
- Waste should be segregated and recycled where practical.
- Care must be taken to prevent spillage of liquids, and spills should be cleaned up and disposed of safely.

Chapter 6 - People

This chapter gives an overview of important industrial relations issues, but I would encourage Project Managers to attend further training to have a better understanding of all principles and requirements.

Project Managers should understand the industrial relations laws of the country where the project is located, their company's policies and procedures, as well as any project specific agreements. Failure to understand and comply with these policies can lead to costly mistakes, and unhappy, unproductive workers.

Case study:

One of the projects I was working on had a project labour agreement with a number of additional rules and payments over and above normal industry practice. One of these was that if a person on an hourly rate worked nine and a half hours, or more, they would be entitled to an additional fifteen minute paid break. The Project Manager did not read this clause properly, believing the additional break only applied if the workers worked more than this, and consequently he set the working day at nine and a half hours.

The project was into its third month when workers queried why they were not being paid for the additional break. After investigation we had to back pay all the workers, which since this break was effectively taken in the overtime period, we had to pay at overtime rates. Overall this cost the project twenty thousand dollars!

From then on, in order not to incur the cost of the additional paid break, we shortened the standard working day by fifteen minutes, which of course caused unhappiness amongst the workers, because they had become used to being paid for nine and a half hours.

On the same contract another item in the agreement stipulated that if a worker used their own transport to travel to the site they were entitled to an additional allowance of twenty-one dollars a day.

Several months into the project, the workers asked to be paid the twenty-one dollars each, for every day on the project. The workers were staying in accommodation provided by the company, ten kilometres from the project, and less than a third of them used their own transport, most shared, in fact, some of the workers didn't even have their own transport. Yet at this late stage, it was impossible to differentiate who had used their own transport and who had shared, since we hadn't monitored them, and when questioned everyone claimed to have used their own. Since it was impossible to dispute their claims we had no choice but to pay every worker the additional allowance from when they started.

If the Project Manager had read the employment conditions properly at the start, and implemented a system of either the company providing the transport, or transport being shared equally between the workers, a considerable amount of money could have been saved.

Of course, the above two examples have a direct cost which can be quantified, but there are also indirect costs not as easily quantified, because any industrial or pay issue results in worker unhappiness. While workers are querying between themselves or asking management what payments they are entitled to, there's often an air of discontent that invades the project, usually resulting in productivity losses. Workers may feel they are being cheated, often with a threat of industrial action if the problem isn't resolved.

The lessons from the above are that the Project Manager must understand the particular labour agreements and conditions for the project, ensure they are complied with, and that the project is administered in accordance with them from the start. These agreements may include rates of pay, special allowances, working hours, disciplinary procedures, travel arrangements, to name a few.

Company policies and procedures

All companies should have a set of policies and procedures in place that governs industrial relations, and the Project Manager must read and understand them, and have them readily available, so they can be referred to. It's important the project adheres to these policies, unless it has been agreed that the rules are different due to special client requirements or local requirements of the state or country.

Project Supervisors and senior site management must be aware of the company's policies and procedures, as well as the project specific conditions. It may be appropriate to arrange training, which would include amongst others; disciplinary and grievance procedures, project labour agreements, the rights of unions, and pay procedures, to enable the management team to better understand them. This would prevent problems created by policies not being applied uniformly across the project by the different Supervisors.

Employment contracts

When employing new workers the Project Manager should ensure they are issued an employment contract and that the conditions in it comply with the legislation of the country, the company's policies and procedures, as well as the specific project conditions and rules. In many companies a human resources department will prepare the contract, but the Project Manager still has the responsibility to review it, and ensure it complies with the project conditions. They are also responsible for ensuring the correct contract is used.

When workers are transferred between different projects, it may be necessary to issue them with a new, project specific employment contract or a secondment contract. Again, this may be prepared by the human resources department, but it's still the Project Manager's responsibility to ensure it aligns with the specific project conditions. The employee should never have terms of employment that are inferior to the legislated conditions, be worse off than the company's basic conditions of employment, nor be paid less than the basic rate of pay stipulated in their existing contract. (It's important to remember that when the employee is transferred from the project he is provided a letter stating that their employment conditions will revert back to those in their original employment contract)

Case study:
On one of my projects the contract document specified that employees returning to work after their rest time at home were required to return to the project on the

last day of leave. However, the employment contract, which the company gave to each employee, specified they were to return to the project on their first working day. Half way through the project the client queried why our personnel were returning to work on their first day of work, and insisted we comply with the project rules. Our staff were unhappy with this instruction, since it resulted in them having less time at home, and this unhappiness impacted on their productivity. We had to get all the staff to sign a revised employment contract for the project, specifying the new condition, and in order to do this, we had to offer them additional compensation. Yet if the original employment contract had been drawn up correctly, in accordance with the contract document, the personnel would probably have accepted this condition without the need for additional compensation.

I have also had occasions, when employees have been transferred to a project and not been given a revised project specific agreement to sign. Consequently some weeks later they were in the Project Manager's office querying why they weren't paid the allowances which others on the project were. This is usually easy to correct, but the employee is often disgruntled and unhappy until the error is rectified, affecting their productivity, and maybe even the productivity of other employees when the disgruntled worker complains to them. Once the complaints and muttering start, there is no knowing what else will come up as a complaint.

Another problem of transferring employees to a project without them first signing a revised contract is they arrive on the project not knowing their conditions of employment have changed. The employee is now already on the project, unhappy with the contract, and there's a possibility they could refuse to sign it, leading to possible industrial relations problems.

What we often overlook is how even a small amount of money affects employees – they can make a major issue of being short paid fifteen minutes, or a losing a fifty cent allowance. To us, as management, the money may seem insignificant, but it shouldn't be underestimated how seriously employees will take the loss, often costing the company more money in downtime while they are in the office complaining.

Personnel records

Personnel records must be maintained for all employees, each should:
- have a personal folder containing:
    - their contact details
    - emergency contact details for their next of kin
    - copies of their training certificates, qualifications, licenses, and visas
    - their employment contract
    - a record of past disciplinary procedures
    - copies of incident and accidents reports involving them (including investigations and medical treatments received)
- be kept current
- kept in a secure place
- be sent to the next project when the employee is transferred, ensuring the records remain secure

- if a person is terminated, contain all the termination paperwork, including any disciplinary documentation, then be sealed and submitted to the company Head Office where they must be secured and stored in accordance with labour regulations

Records of all time sheets and payslips should be available on the project in case they need to be reviewed, or if employees have queries. Sometimes employees query these records several months, or even years, later.

Identity documents, passports, visas, permits and qualifications

The Project Manager must check employed personnel are legally permitted to work in the jurisdiction, state, or country, where the project is located. This compliance includes ensuring they have a legal form of identification, and the required permits and visas which allow them to work on the project. Some projects may also require personnel to have particular qualifications or trade certificates, and a copy of these should be kept on file.

When the project is located in a foreign country there are usually restrictions on the importing of personnel, with those brought in from outside the country requiring valid passports, and meeting the requirements to enter and work in the country. Failure to meet these requirements could result in fines to both the company and the Project Manager, with the personnel being removed from the project, and possibly from the country.

Some projects and clients may also require personnel to have police clearances or medical certificates before they can be employed on the project.

Discipline

Discipline must be applied in a fair, even and consistent manner. The standards should be set from the start of the project, since it's difficult to later enforce rules which have previously been overlooked.

The company rules should be included in the employee's contract, and then covered in the project specific induction discussed in Chapter 2.

It's essential that employees understand the disciplinary procedures so there's no excuse for unacceptable behaviour and non-compliance with the site rules.

Disciplinary transgressions must be dealt with in accordance with the company and legislated procedures. On occasion, I've had a project not follow the disciplinary procedures correctly resulting in a dismissed employee having to be re-employed. This employee has to be reimbursed their lost wages, in addition there are the legal costs in defending the dismissal, and apart from which, re-employing a dismissed employee undermines the discipline on a project.

Lead by example

Employees usually notice what Project Managers do, and if they are repeatedly absent, take extended lunch breaks, arrive late, or leave early staff may think this is acceptable practice and do the same thing. When workers are reprimanded, they often bring up the subject of when the Project Manager appeared to break the rules as an excuse for their behaviour. For this reason the Project Manager must ensure they comply with safety rules and the behavioural code of conduct for the project.

Drugs and alcohol

Drugs and excessive alcohol consumption can be a major problem on a project leading to:
- accidents and incidents
- violence
- poor behaviour
- poor work performance
- absenteeism

To ensure that alcohol and drugs don't become a problem, ensure:
- rules are in place and enforced from the start of the project
- there is a policy of zero tolerance
- each employee on the project understands the limits, these should be included in their employment contracts
- the project has a calibrated breathalyser, which is used regularly to test personnel

Case study:
We had one project where there were numerous instances of people arriving late for work, being absent without leave, as well as numerous safety incidents. Many of these incidents were probably related to the excessive consumption of alcohol after-hours, and, although there was a breathalyser on site, the tests were not administered correctly with few instances of people being found over the limit. Yet, as soon as we put a trained, responsible, person in charge of these tests, the number of staff found to be over the limit at first increased, but once employees realised they would be caught if they were over the limit, their numbers decreased and productivity, attendance and safety improved.

Friends and relations

Many projects employ friends or relatives of personnel already employed on the project. This has to be handled with care, with no signs of favouritism or bias towards them, because other employees will notice if they are treated differently. If a friend or relative does not perform, or transgresses a site rule, they must be treated in accordance with the disciplinary procedures, the same as any other employee would be.

Sometimes the client, or a member of their team, requests the contractor employ a member of their family. This request should be dealt with carefully, since it can result in a conflict with the person who made the request, especially if the relative employed is a poor worker, transgresses the project disciplinary code, requires disciplining, or even has to be dismissed.

Indigenous and local employment

Many projects require that a certain number of local and indigenous employees are employed – it's important the project complies with this quota.

Often the project is located in a rural area, with high unemployment or low wages, and the locals have an expectation that they will be able to find well-paid employment on the new construction project. If the contractor arrives, bringing their large equipment and own employees they will be disappointed, and enviously watch these well-fed, clothed, and relatively affluent employees, passing through their community, while they remain unemployed. This can lead to anger and unrest, which can turn to theft and even disruption of the project.

Contractors must be mindful that they should be seen to be supporting the local community, and the best way of doing this is by providing jobs and training, even if the locals appear to be uneducated, unsophisticated and not used to working on a modern construction site. If contractors employ local people, train them to become productive employees, hopefully this training and experience will enable them to gain employment elsewhere after the project is completed. I've often employed local workers who have proved willing, and become excellent workers, and were then transferred to our next project. Although the training can sometimes be costly and time consuming, there are financial benefits, locals often don't require the transport and accommodation which may have to be provided to employees brought from outside the area.

We all owe it to develop the indigenous population, however sometimes this can take a great effort since we need to understand their cultures so as not to cause misunderstandings and conflict.

When employing local or indigenous people it's important staff understand the necessity for employing, training and retaining them, because if Supervisors are not committed to this, it could become a problem, leading to the indigenous employees leaving the project. There are some Supervisors who are better at training, developing and working with indigenous workers than others, and where possible indigenous employees should be placed with them.

Cultures, ethnicity and backgrounds

A construction site is usually a melting pot of people, from different cultures and backgrounds. The Project Manager should have an understanding of the differences on the project, and be sensitive to these in their communications. It's important to understand some of the vagaries of these since people from certain cultures may take offence at things that are commonly said, done and accepted in your culture, and you may take offence to things they do, which may be regarded as normal in their culture.

Understanding cultures and backgrounds can also be useful when negotiating, and different negotiating techniques can be used depending on the person's background. One technique may yield little or no success with one individual, while the same technique can be used successfully with others.

Understanding the cultures and backgrounds of people working on the project will enable better relationships and teams to be formed.

Training

Training is an important aspect of developing employees and ensuring the project has the required skills to carry out the work. Training can take many forms, for example:
- it can simply be on the job training and mentoring given by Supervisors or other suitable personnel
- workers on site could be gathered together and shown how to do a

particular task
- formal training could be provided by a specialist provider brought to the site
- the employee may attend a formal training course off site

Where possible:
- training courses should be given by accredited training providers
- a certificate of attendance, or competency, should be given to the person who successfully attended, and a copy should be retained in the employee record folders
- training shouldn't be done just for the sake of training, since after a person is trained, they will expect to be employed in that role, with their wages adjusted accordingly, so if there aren't employment opportunities in that field you should not be training people for the role
- projects should maintain a training matrix which should mean:
    - monitoring takes place to ensure the training data is kept up to date
    - personnel are competent to perform the tasks
    - competencies are kept valid
    - as new employees arrive their details are added to the matrix
    - when they leave their details are removed

Feedback

It's important you provide both positive and negative feedback to your staff. I often have Project Managers complaining about the quality of a person, even saying 'they are incompetent', however on most occasions when I enquired if they've spoken to the person and explained their short comings, they normally haven't. In fact, the problem person often thinks they are doing a good job. It's therefore important, to tell a person when they have done something incorrectly, or aren't performing according to expectations. If you do this you may find a dramatic improvement in their performance.

Poor performance may also be related to the person having insufficient knowledge to perform the task, so their performance may improve if they are given additional training or coaching. Certain people are also better at certain jobs than others, and if the right niche is found they may perform well.

Give feedback in such a way that:
- it doesn't always appear to be negative and critical
- negative feedback isn't presented in a public place in front of other staff and workers
- it isn't shouted or given in a rude or abrupt way
- the problem is explained with a suggestion on how you think they can improve their performance
- Supervisors should be involved when one of their workers is praised or criticised

If a person's performance doesn't improve, and they can't give a reasonable explanation of why they will not, or cannot, improve then consideration must be given to following the disciplinary process.

People appreciate being thanked for their efforts, and positive feedback should be provided if a task is done well, a milestone is met, quality standards are exceeded, or a task is done safely. This praise can be public, but not an everyday event or it will appear that you praise anyone and everyone, and it loses its impact.

Grievance procedures

All employees should understand, and follow, the grievance procedures, which are in place to ensure matters can be dealt with swiftly, fairly, and the issues that lead to them can be resolved to prevent further problems. When employees don't follow these procedures, often because they are unaware of them, the problem is not attended to or resolved, causing discontent, which may impact on productivity, even leading to labour unrest.

The management team should also be aware of these procedures, so they can ensure the employees follow the proper process, and the problem is directed to the correct authority ensuring its resolution.

High labour turnover

If the project has an excessively high rate of employee turnover the Project Manager should ascertain the possible reasons for this, since a high turnover rate:
- is disruptive to the project
- causes the project to lose a skill which has to be replaced which could take some time
- can cause a problem when there's a shortage of skills to replace the person leaving
- means each time a new employee arrives on site they have to learn the project rules, which may take time before they become effective, resulting in lost productivity
- increases recruitment costs
- creates additional employment expenses, such as medicals, inductions and personal protective clothing
- sometimes requires the replacement comes at a higher salary
- often sees the best employees leave first

The appropriate way to ascertain the reason for a person leaving is to hold an exit interview with them.

There are many reasons why people resign, and sometimes the problem is simple and easy to remedy for example:
- it may be due to a particular staff member being abusive
- sometimes employees resign because there are more attractive working opportunities elsewhere, and you may not be able to do much, but it could be worthwhile to investigate what changes can be made to make the project conditions more attractive
- sometimes employees are 'poached' by other companies working on the project, in which case the Project Manager should raise the issue with the client, or managing contractor, so they can put a stop to this because it's detrimental to the overall project progress and costs the affected

contractor money

When the Project Manager talks to an employee who has resigned, they may find the employee hasn't thought through the reason for leaving as fully as they should have. The employee may in fact be leaving for what they only perceive to be a better deal, but when their current conditions of employment are compared with their new ones, they may find they will be worse off and choose to withdraw their resignation.

Leave

This may seem a strange statement, but do ensure you take your full allocation of leave. I think I've been happiest, and consequently probably performed my best at times when I was planning my leave, about to take leave, or just returned from leave. I have taken leave often, and yet all my projects went on normally in my absence. Nobody is completely indispensable. The break doesn't have to be long, maybe only an extended weekend, but it should be far from work. Obviously try and plan the leave around the project requirements, although there won't be a perfect opportunity. It will be viewed poorly if you request leave for a period that includes a critical hand-over date, or when a tender you're responsible for is due. Also, you shouldn't expect to take more leave than your entitlement.

The second item links to the first, in that staff should be allowed to take leave. I wasn't always the most gracious person when someone in my team put in a leave request, in fact I was probably quite grumpy, and made a sarcastic comment, however I seldom refused a leave request unless it was totally unreasonable, or their leave entitlement was exceeded. When a member of your staff takes leave it will probably put an extra strain on you and the rest of the team, but hopefully it will create a happier employee who will perform better on their return.

Ensure employees are not working weekends or additional overtime to bank this extra time worked and take it as leave later. It's counterproductive since the person working day-after-day, without taking their rest days off, will become inefficient and their performance will suffer.

Leave should be recorded on a leave request form, which should be approved by the Project Manager before it's submitted to the payroll office. All salaried staff should complete a leave form if they are absent from work, whether its paid annual leave, unpaid leave, sick leave, study leave or family responsibility leave. Where necessary, such as sick leave, supporting documentation should be attached. Failure to submit a leave form results in employees being paid for time they didn't work, and possibly even in excessive absenteeism.

Gifts, bribes and corruption

There should be a clear policy discouraging bribery, corruption and the receipt of gifts. The policy should be displayed on the project notice board. Staff should inform the Project Manager of gifts they receive from suppliers or subcontractors, as well as invitations to attend sporting events or functions, because when someone receives a reward from a supplier, or subcontractor, they may be tempted to deal leniently with that company if there are claims, or if the company delivers a substandard product.

Under no circumstance should bribes be paid to anyone. I know, in some countries, bribery is almost second nature to the local officials, however, it sets a precedent for all future interactions, and becomes very expensive. Even though local officials may expect such gifts there will always be laws prohibiting the practice, and you will be guilty of breaking them should you pay a bribe, and possibly face a fine or even prison. The paying of bribes is also actively discouraged by many clients, government agencies, and aid organisations, and if they discover a company has been implicated in paying one the company may be prevented from ever working for them again.

It's the Project Manager's responsibility to ensure all project staff understand the consequences of paying or accepting bribes, because it could lead to serious consequences for the company and the Project Manager.

If the Project Manager receives a gift or invitation from a supplier or subcontractor, they should confirm with their manager whether they can accept it.

Before the contractor gives the client, and their representatives, a Christmas gift, or invitation to a function or sporting event, the Project Manager should ascertain the client's policies regarding the receipt of gifts. This way nobody is placed in an awkward position.

Functions attended by staff from the contractor and client can be useful in fostering better working relations, however, I have seen these become too frequent.

Unions

The Project Manager should have a copy of the agreement the company, or project, has with a union. This agreement usually governs the employment agreement, and may regulate what tasks can be done by what category and trade of employee.

The unions have certain rights of access to the project, however, they do have to comply with the relevant notifications and regulations that govern this right. The Project Manager should ensure the union representatives are treated cordially, and respectfully, but they must also ensure the union representatives abide by the rules governing their access.

The art of persuasion

Project Managers sometimes have people reporting to them who are older, more experienced, and often used to performing tasks in a certain manner as they have countless times in the past. They usually don't want to behave differently, even though there may be a safer or more efficient way. A direct order to do the task in another way will often result in resistance and an unhappy member of staff, consequently the Project Manager has to be able to persuade the person that the alternative method is beneficial, and convince them to proceed as suggested.

Project Managers skilled in the art of persuasion are successful since they can easily convince staff and clients to follow their suggestions, and decisions, while maintaining harmony on the project. There are effective courses which will train and guide people in the art of persuasion.

Conflict resolution

Conflict resolution is an important aspect of a Project Manager's duty and potential conflict situations should be dealt with, and resolved, in a fair and impartial manner, otherwise trivial matters can quickly escalate in to major issues.

Case study:
   I had an Engineer take offence with me because he felt I ignored an issue he had with his Project Manager. Then because the Engineer was upset with me, his performance and productivity was affected to the detriment of the project.

Summary

- The Project Manager must understand:
    - the project specific labour agreement
    - the company's industrial relations policy
    - the legal requirements relating to the labour laws of the country
- It's advisable to consult with an industrial relations expert to ensure the project is in compliance with these policies and agreements.
- The project staff must be aware of these policies and procedures, and given further training to better understand them.
- The aim of the policies and procedures is to ensure labour harmony on the project, which results in safer working conditions and a more productive and stable work force.
- Personnel should sign a contract of employment.
- The conditions of employment may vary from project to project.
- When employees are transferred they may have to be issued with a revised, project-specific employment contract.
- The contract of employment must comply with the specific project conditions, as well as existing industrial relations policies and agreements.
- Discipline must be applied in an even, fair, and consistent manner, from the start of the project.
- The Project Manager must be able to:
    - deal with grievances
    - resolve conflicts
    - negotiate and persuade
    - deal with people of varying levels of experience, knowledge and education, who may come from different cultural, and ethnic backgrounds
    - give feedback to employees
    - lead by example
- Appropriate training and development of personnel is important.
- The employment, and upliftment, of the local and indigenous communities should be encouraged.
- Substance abuse should not be tolerated.
- Bribery and corruption must be discouraged.
- Reasons for high labour turnover should be investigated.

Chapter 7 - Plant and Equipment

In this chapter I'm referring to plant and equipment required to construct the project, not the equipment that's built into the project to become part of the final facility. Plant or equipment may only form a small part of a project, or in the case of earthworks it could be a major component. It includes small tools, generators, compressors, buses, utility vehicles, scaffolding and formwork, all the way up to major items of earthmoving equipment and heavy-lift cranes.

I don't intend to cover the technical side of what the different and appropriate uses of the equipment are, but rather focus on some of the costly and hazardous pitfalls that occur in administering the plant and equipment.

The plant and equipment may be hired from within the company (many companies have internal plant divisions or departments), from an external hire company, or bought specifically for the project.

The right item for the job

Use the right equipment for the job! How basic could this be? Yet this rule is often not followed, resulting in inefficiencies, damage to equipment, and in the worst case injury to the workers. In fact, we are probably all guilty of using the wrong equipment. How often have we used a knife as a screwdriver, or a shoe or piece of wood as a hammer?

Before ordering the item of equipment, the contractor should know what the item will be used for, and the production required from the machine. Research the types of equipment available and their capabilities. The uses and expected outputs of equipment can be obtained by talking to the sales, or hire, representatives, however, don't believe everything you're told, since some may not be as well-informed as they should be. Production rates are also available on the internet, but rather, where possible, talk to someone who has used the item for a similar purpose.

Just as important as it is to choose the right size bulldozer or excavator, it's equally important to choose the correct electrical drill or saw, because often, contractors choose power tools meant for household jobs instead of major building renovations. An underpowered, light machine cannot withstand the heavy usage of a building site and will break down, while its inadequate power slows the task down, leading to frustration, inefficiency, and sometimes even injury.

Some golden rules to remember are biggest isn't always best, although rather err on the larger size than the undersized, and, cheapest is also not always the best.

Internal hire, external hire or purchase

Whether an item of equipment is hired internally or externally is a decision often taken out of the Project Manager's hands, and is usually dependant on the company. If the company does not own any equipment, then obviously internal hire is not an option. However, should the company have the item, and it's not being used elsewhere, the company will want it used on the project. However, on remote sites, when the item is only required for a short period of time, and one is available close to the project it may be more cost effective to hire it externally.

Sometimes the cost of hiring the item externally is cheaper than internal hire, and in this case there is always the temptation to hire it externally to save the project money. However, the Project Manager should be mindful of the bigger picture, which is not just about one project making a profit, but the company as a whole. After all it would be pointless if a project is paying money to an external hire company, when a similar item, owned by the company, is standing idle elsewhere. The project may save a few dollars but overall it's costing the company additional money.

An alternative to hiring externally, which Project Managers don't always consider, is for the project to purchase the item. I've often found hired items that have been on the project so long, that the hire costs far exceeded the purchase costs.

So before hiring a machine externally thought should be given to how long it's required. Cost estimates and comparisons can then be done, comparing the cost of hire for the duration, versus the purchase price, factoring in the costs of repairs and maintenance which the project will carry for the purchased item.

External hire

When hiring externally, enquire about the age and condition of the item, because often an item is available at a cheaper rate because it's old. Older machines:

- may not perform as efficiently as newer machines since they have lost some of their original power, resulting in a loss of productivity, which may even mean the machine is unable to do the required task
- frequently break down resulting in lost time and production, which also affects the morale of staff and operators
- are often not as fuel efficient as newer models
- may not have the safety features of a new machine, and consequently may not comply with the client's safety requirements
- are more likely to have oil leaks or burst hydraulic hoses, which create environmental problems
- may be noisier
- may have more visible smoke pollution
- may be a safety risk with failures, resulting in the item catching fire, or moving out of control (I've had the brakes on machines fail, resulting in them running out of control, almost causing serious accidents)
- are possibly less operator-friendly, with a poorer ergonomic design, causing operator fatigue, resulting in loss of productivity

It should also be noted that the client often views the plant and equipment the contractor has on site as an extension of the company, even though they may be externally hired. Therefore old equipment, which frequently breaks down, creates a poor impression of the company. Equally so I've have had projects where the hire company supplied new machines, and the client complimented us on our new equipment, even though it was obvious it was externally hired.

Sometimes, however, if the machine is not being used for a critical task, nor is it going to be working at full production, consideration can be given to hiring an older, cheaper machine.

Of course, the external appearance of a machine does not always give the true picture. On many occasions I have had machines delivered to my project, looking good in a shiny coat of paint, and yet, be in such a poor state of repair they couldn't be driven off the delivery vehicle. It pays to check the year of manufacture, the hour meter reading, and to request the service and repair records, since this will give an indication of how old the item is and how well it's been maintained.

Ordering equipment

When ordering equipment:
- ensure the machine is suitable for the task
- inform the hire company what the machine will be used for, as this may affect the hire cost, or which attachments are sent with the machine (this is particularly relevant when ordering excavators which come with a variety of different buckets)
- the hire company must be aware of the specific project requirements for the item of plant, which may include amongst other things, reverse hooters, flashing lights and fire extinguishers
- there may be documents and paperwork required before the equipment can operate on the project, such as maintenance records and test certificates (I've had items of equipment parked at the project gates, unable to work because the correct documentation was not supplied, this affects production and results in paying for the operator and machine while the machine can't work),
- specify the duration the item is required for, since often the longer the item is required the more competitive the rate, although the project may be penalised if the item is returned to the hire company sooner than that specified in the hire agreement
- ensure the item will be available for the required duration

Case study:

On one of my earthmoving projects we ordered the equipment for the original duration of the project, however, during the course of the contract the amount of ground we had to move was doubled by the client, in addition we were achieving a lower rate of production than anticipated. This meant we required the machines for a much longer duration than was originally expected, but the hire company refused to extend the period of hire for their machines since they were required elsewhere. Consequently at the end of the original hire period they removed their machines, which meant we had to obtain other machines to complete the work, causing us additional costs and disrupting the production.

Transporting the equipment

When organising the transport of equipment:
- advise the hire company who will be transporting the equipment and when they can be expected
- ensure the item will be ready (I cannot tell you how often I've heard of transport arriving to collect an item of equipment, only to find it's not ready, resulting in the project incurring standing time costs from the transport company, who in some cases may be unwilling, or unable, to wait until the item is ready, resulting in alternative transport having to be organised causing delays and additional costs)

- ensure the required permits are in place, which sometimes can be numerous depending on the route the item has to be transported over
- ensure there are no restrictions on the project site which may prevent the load from being delivered (sometimes they may be of a temporary nature, like the client doing work on the access road, or another major delivery already scheduled for that day)
- check the item of equipment is covered by insurance while it's in transit
- it may be prudent to check where the transport company intends to park the vehicle should they have to stop-off at night en-route to the project, because in some regions and countries it may be unsafe for the vehicle to overnight on the side of the road where the equipment can be vandalised
- the transport company must be made aware of the project conditions of entry, which may include the condition of the vehicle, and its compliance with the project safety rules
- the transport company must be given a clear map and directions of how to get to the project, and contact details of a person on the project, should they get lost, need assistance or need to be met on arrival
- project site staff must expect the item and know where it should be offloaded (many construction sites cover large areas, and a heavy piece of earthmoving equipment offloaded at the wrong place can require special transport to move it to the correct location, which could cost several thousand dollars, and cause several days delay)
- ensure the equipment can be offloaded at the project, which means the item should arrive within normal working hours
- on certain projects the item may have to be inspected by a representative from the client before it's offloaded, so ensure this person will be available
- the correct paperwork for the item should be available, without this they may not be authorised to be offloaded
- ensure there is a risk assessment for the offloading, and even if this isn't a project requirement it should be in place
- a qualified person must be available to offload the item

Case study

On one of my first projects, before we had started any work, we took delivery of a large excavator which arrived on a low-loader transport vehicle. The operator of the machine decided to walk the machine off the back of the low-loader since there were no ramps, and it was normal practice to save time. On this occasion the driver had parked the vehicle on a slight slope, with the deck tilted to one side. As the operator manoeuvred the excavator to the back of the low loader, the excavator's steel tracks had no traction on the inclined steel deck so slipped to one side, it then fell off, landing on its side on the ground. Thankfully when the excavator began slipping the operator jumped from it, fortunately exiting on the opposite side to the direction it fell.

What a way for us to start a project! We were extremely lucky no one was injured in the accident, the operator or a bystander could have been crushed under the excavator. The excavator had to be loaded onto a recovery vehicle and a replacement had to be brought to site. It was very embarrassing to face the client *with a major accident before we had even started work.*

Lessons from this are not to take the offloading of plant and equipment lightly, to ensure competent and trained personnel supervise and undertake the offloading, and that an appropriate and relevant risk assessment is in place.

Just as important as offloading the item is considering how the equipment will be loaded and transported off the project when it's no longer required. This may sound simple, but it should be remembered the project will change with time, and what is now empty space may eventually be occupied with new buildings and structures, or blocked by other contractors. This is particularly relevant in congested building projects in the city.

It could also become impossible to get large trucks, or the cranes required to load the item onto the project site. There may be new access roads or bridges, which aren't designed to carry the item of equipment, or there may be new overhead structures, such as gantries or pipe racks that restrict the height of vehicles passing under them. These obstacles could result in additional costs to remove larger items, such as; additional hire for the item to remain on site until suitable access is available, costs to dismantle the item, or the cost of a larger crane to load it.

The safety of the loading operation is as important as the safety when offloading it. In fact I've experienced a number of accidents when trucks were being loaded at the end of a project because it's often done hurriedly, people have become complacent, and those more familiar with the safety requirements have already demobilised from the project.

Case study:
One building project our company completed used a tower crane to construct the structure, when this was completed the crane, which was no longer required, was dismantled. Taking-down a tower crane is always a dangerous operation so it's done by experienced professionals, in this case the same team that had previously removed dozens of cranes for our company.

Once the crane was safely dismantled its components, which are fairly long and bulky steel sections, were loaded onto the waiting transport trucks using a mobile crane. Two sections were stacked on top of each other, on each truck, and tied down. The operation proceeded smoothly and was almost complete when the crane loaded a second section onto a truck. This was unhooked, and personnel began to tie the load to the truck when suddenly the section slipped off the side of the truck – onto the Supervisor – killing him.

When the equipment arrives on site

The task of receiving equipment deliveries is often left to a person who has not been told what to look for. Things to check include:
- is the item actually for the project (sounds very basic, but it wouldn't be the first time something is delivered to the incorrect project)
- is it the item that was ordered
- damage, which should be noted on the delivery note, put in writing to the supplier, and if possible photographs should be submitted as well with a set

kept on the project, so as to avoid being charged for the damage when the item is returned
- the condition of wearing parts and cutting edges since the project is normally charged for their replacement
- the condition of the tyres, because those that are damaged or in a poor condition often result in deflated tyres, which not only cost time to repair but the project could be liable for new tyres when the item is returned
- ensuring it is in a roadworthy or in an operational condition
- that all the parts and fittings are included with the delivery, and that wheeled machines come with at least one spare tyre
- the fuel level of the machine which should be recorded since suppliers normally expect their machines to be returned full of fuel, and will charge for additional fuel they have to put in (but of course the converse doesn't happen, and often machines arrive with minimal fuel, which when large items take several hundred litres of fuel, could result in a significant cost)

Most importantly the project member who requested the item should be notified the item has arrived on the project. Often machines arrive and then aren't used for several hours, or days, because the person who requested the machine doesn't know it's available.

Understanding the conditions in the hire agreement

Often items of equipment are hired but project staff have not read the terms of the hire agreement, which can be fairly stringent. The terms should include:
- the rate for the machine and whether it's charged hourly, daily or monthly
- the minimum hours or minimum days the item will be invoiced for
- whether a surcharge applies if the machine is worked excessive hours or travels more kilometres
- what the hire charges will be if the machine can't work due to rain, force majeure, or non-work days on the project
- which party is responsible for insurance
- who is responsible for repairs and maintenance (these costs can sometimes be large, adding significantly to the hire costs)
- who supplies fuel, oil and cutting edges, and for cranes the slings

It may be possible to negotiate some of these terms when the order is placed, which could result in considerable savings later.

Personnel using the machine must be aware of these conditions. In the past I've had Supervisors park-up machines thinking they are saving hire costs, when in fact the machine is still incurring charges since it hasn't worked its specified minimum hours. However, by swopping machines around, and using it on a task more aligned with the hire agreement, can sometimes result in some hire costs being saved.

Booking of plant hours

This is a task that should be simple, and yet many projects seldom administer it correctly.

It's important that:
- a responsible member of staff is tasked with signing the time sheet for the

Chapter 7 - Plant and Equipment | 127

machine (all too often, the signing of a machine's time sheets is a task left to a leading-hand or operator who may not check the hours properly)
- the person signing understands the implications of what they are actually signing for, which is the record for payment purposes
- the person signing the time sheet understands the conditions of the hire agreement, for instance are there minimum hours that the machine must work in a day, is a lunch break included in the paid hours, and what happens when it rains
- the time sheet must be signed off on a daily basis so as to avoid misunderstandings, or arguments, later over the hours worked (I've witnessed on occasion where this is only done at the end of the week, the end of the month, and even at the end of the contract)
- the start and finish times are recorded, as well as the times of all breaks
- the time the machine was broken is recorded, and if possible note the problem

Care and maintenance

Equipment and plant must be cared for and maintained properly since damage is costly to repair and often results in down time and lost production. This care includes:
- the operator:
    - should if possible be dedicated to a specific machine
    - should be certified or licensed to operate the machine
    - must be competent to operate the machine (even if the hire company has provided their own operator, verify the operator is competent, and has the required current licenses for the machine, I've often had plant hire companies send their machines to site with unlicensed operators, or operators who weren't competent, causing a safety risk, adversely affecting production from the machine, and causing damage to the machine and other property)
- the machine being operated safely so that it does not collide with other machines or structures
- it being used in areas where it will not overturn, become bogged, or damaged by falling objects
- ensuring the equipment is not used to perform tasks it was not designed to do
- ensuring equipment is not overloaded
- operators doing prestart checks on equipment at the start of the shift, or when they use the equipment for the first time each day, and:
    - reporting any faults and damage detected in the course of these inspections
    - recording checks on a prestart checklist
    - completing inspections which include checking the fuel and lubricant levels, and the condition of brakes, tyres and wearing parts
    - tagging the machine if it's unsafe to use until an authorised person has rectified the fault

- immediately reporting and investigating any damage or incident involving the machine
- regularly cleaning the machine
- recording all routine maintenance, work done during the maintenance and any repairs done on the machine
- using the correct parts for all repair and maintenance work
- the Project Manager, in passing, inspecting the condition of the machines on the project to ensure they are clean, undamaged and that prestart inspections have been completed

Service records

Keeping a record of the work done is as important as doing the maintenance itself because it enables the next maintenance to be planned at the correct interval.

Service records are also vital if there is an incident or accident involving the item. If the item was serviced by a qualified person, on the due date, and this was recorded, then if there is an incident resulting from a failure of a part on the equipment, the contractor and Project Manager should be protected. However, if no record can be found, it will be difficult to prove the failure was not a result of poor maintenance, which could have serious consequences for the Project Manager, particularly in the case of a serious accident.

Calibration records

Some equipment requires calibration at regular intervals, as well as after it's repaired or serviced. Projects should maintain a register of equipment requiring calibration, since items that aren't calibrated correctly (for instance survey instruments and testing equipment) can lead to the incorrect readings which can in turn result in a serious error.

Personnel using these items should be aware they must be treated with care, and if there are concerns the item is faulty, it should be checked to verify that it's still providing accurate readings. If there are any further doubts it must be sent for servicing and recalibration immediately.

Protection of equipment

The equipment and machines should be secured when not in use. For small power tools this means locking them in tool boxes or returning them to the store at night. Large items of equipment should be parked safely in secure areas where they won't get damaged by passing vehicles, be vandalised, or have parts stolen off them.

Items should be locked and their keys placed in a secure key cabinet in the office. On numerous occasions I've encountered machines parked on a project with the keys left in them, resulting in the machine being used by an unauthorised person, without the correct license, and in some cases leading to property damage. I've also had trucks and vehicles stolen from the project for the same reason.

What can also happen is that operators take the keys home with them at the end of the shift. Then they either return to the project the next day forgetting the key, or are sick and don't come to work, and the machine can't be moved or used until the keys are located.

Overloading of equipment

Overloading of equipment not only causes damage to the equipment, but can also result in accidents. In the case of overloaded trucks it can result in overturning, damage to tyres and suspension, or objects falling off the back causing injury. Also equipment that's underpowered for a task, or used inappropriately, will wear out sooner and may cause injury should a mechanical part become overstressed and break.

Reporting breakdowns

Often breakdowns of hired equipment are not reported or are recorded incorrectly, resulting in the item going unrepaired for an extended period causing lost production, as well as the project incurring hire costs for an item which cannot be used. It's therefore important that staff understand, and follow the appropriate procedures. These should be that:
- the breakdown is reported immediately to the contractor's site management
- the breakdown is reported immediately to the hire company
- verbal notification to the hire company must be followed in writing (via fax or email) and should include:
    - the date and time of the breakdown
    - the item of equipment including its number
    - what the problem is
    - who the breakdown was first reported to and when
    - the name and contact details of the person reporting the breakdown
- the Contract Administrator receives a copy of this notification to ensure hire is not paid while the item is broken

If an item is broken and returned to the supplier for repairs, it should be accompanied with an off-hire note and a note explaining the problem. On occasion I've had projects return an item for repair with no documentation, and sometimes the supplier doesn't even know where the item has come from, or even that the machine is supposed to be repaired and returned to the project.

Operator training

It's good practice to have certified trainers visit the project to verify the competence of operators. Making use of a trainer who is an experienced operator to improve the skills of your operators can also be invaluable. These trainers often understand what the machines are capable of, and can coach operators on how to operate their machines safely, maximising the output, without damaging the machine. Part of the training should also focus on the checks and inspections that ensure the machines are kept in a safe and good working condition. The benefits achieved of this will outweigh the costs of the training.

Formwork and scaffolding

Many projects require tons of scaffolding and formwork equipment, but there are many varieties, and their costs can vary considerably, so choosing a particular system can be quite confusing. (Refer to Chapter 1)

The formwork and scaffolding should be checked when it's delivered and before each use to ensure it's in good condition, because failure of a component under load can result in a serious accident. Damaged formwork may also result in the completed structure having a poor quality finish. The equipment must also be handled with care when it is used, and stacked neatly and securely when not in use, with concrete spillage cleaned off it regularly.

Scaffolding and support-work must be built by trained personnel. Once erected it should be checked and signed off by a competent authorized person. Scaffold 'Safe to Use' tags should then be attached to the scaffold. The scaffold should be reinspected after any severe weather event, when the scaffold is altered in any way, if there are concerns about the safety of the scaffold, and at regular intervals which are normally once a week.

Theft of formwork and scaffolding can be a major problem, particularly when there are multiple contractors using similar equipment, therefore it may be necessary to clearly mark it so it's identifiable, and doesn't get mixed with other contractors' equipment.

Formwork and scaffolding design

Formwork and support work should be designed to ensure it will be strong enough to withstand and carry the weight of wet concrete, reinforcing, cast-in items, and the people and equipment used to place the concrete. The design normally assumes a rate of pour, and if this rate is exceeded it could lead to overloading of the formwork. Also materials should not be heaped on formwork, or scaffolding, since this could result in the equipment carrying a load it was not designed for.

Case study:

I was the Project Director on a project that involved constructing a reinforced concrete slab, two metres thick, and eight metres above an operational railway line which the supporting scaffold had to span. The support-work was designed by our Formwork Design Engineer, who also produced detailed drawings of how the support-work should be constructed. The quantity of equipment was ordered from these drawings.

There was a full-time Project Manager allocated to the project, and at the time, I was responsible for five other projects, situated hundreds of kilometres apart, so I visited the project approximately every two weeks. The day before the concrete slab was due to be poured I visited another project, and at the last minute decided to make a detour on my homeward journey to check the slab was ready for the concrete.

On arriving on site I was informed the slab was ready to receive concrete, however, I noticed a number of steel beams that had been ordered were lying unused on the ground. I queried this, and was informed the Supervisor had decided the beams were superfluous and omitted them. I contacted the Formwork Design Engineer, and explained my concerns. He immediately checked his calculations, and called me back to say he had major concerns, and was on his way to the site – some 200 kilometres from his office.

When our Engineer arrived and inspected the support-work, he found the omission of the support beams meant the load of the wet concrete on the slab would not have been distributed evenly, and would have resulted in some of the scaffold support legs carrying double their load capacities. To compound the error, the Engineer found a mistake with his original calculations, which would have anyway resulted in the legs carrying an extra 20% over their capacity. This additional 20%, together with the doubling of the load, would have resulted in some scaffold legs carrying nearly two and a half times the load they were capable of carrying.

If we had gone ahead with placing the concrete on the slab there is no doubt the support-work would have failed with catastrophic results, people would probably have been killed and others seriously injured, the railway would have been closed for weeks, the project would have been delayed by months, and there would have been costly property damage.

As a result we postponed the concrete pour until we were able to install additional support-work to carry the load, ensuring none of the scaffold was over-loaded.

There are several lessons from the above:
- the formwork design must be double-checked to ensure there are no errors or erroneous assumptions
- the design drawings should be followed and Supervisors should not omit items, no matter how unnecessary they may seem, unless they first check with the Designer
- it's important to check that the support-work and formwork has been erected according to the design drawings. These checks should also include:
    - ensuring all bracing is installed
    - all support-work is erected straight, vertical and level
    - the equipment is in good condition
    - the supporting ground is firm and capable of carrying the load
    - the support-work and formwork is all tightened and secured properly
    - remembering that support-work erected on top of existing elevated slabs transfer the load on to these slabs, so they must be designed, and supported in such a way that they can carry this load

Cranes, lifting equipment and slings

Incorrect use of cranes and lifting equipment are often the cause of serious accidents. Lifting equipment must:
- have outriggers properly deployed
- not be used to lift loads that are heavier, or at a greater reach, than their rated capacity
- not have outriggers near the edges of excavations
- have the ground conditions checked to ensure that it can support the maximum loads exerted on the outriggers
- not be used when there is excessive, or gusty, wind
- not be used in lightning conditions
- not be used in locations where one crane can come into contact with

another since contact between cranes may cause overturning of one or other of them
- have trained riggers to control lifts
- have a proper rigging study prepared for heavy lifts to ensure it can be performed safely

The items being lifted by cranes should be properly secured to ensure the load does not shift or become dislodged, while being lifted. Falling objects can not only result in injury or death, but also cause damage to structures, equipment below, and the object itself. In addition, when a load shifts it may become unbalanced, causing the lifting equipment to overturn.

Slings connected inappropriately often result in them being overstressed, causing failure, with catastrophic results. Slings should be:
- checked regularly for damage and damaged slings must be clearly marked as such and removed from the project
- the correct capacity for the weight of the load
- attached to secure lifting points
- in the case of multi-leg slings, be fixed in such a way that the legs are as close to vertical as possible and not be at an angle greater than thirty degrees from the vertical, since the more horizontal the leg is the greater the stress created in the sling
- used with lifting beams where necessary (these should be designed for the task and manufactured by licensed fabricators)

Returning equipment off hire

When a machine has completed the tasks on site, and is ready to be returned to the hire company there are a few important rules that should be followed:
- ensure that the machine has, in fact, completed all the work on the project that it's required for (sometimes I've had projects send a machine off site, only to find a few days later that the machine should have done another task, resulting in a costly exercise to bring the machine back), so it's important that Supervisors communicate with each other, particularly at the regular staff meetings, and advise in advance when they will have finished using a particular machine and what their future requirements will be
- the hire company should be advised in writing that the machine is off hire, and this instruction should include:
  - the names of the company and project
  - the item of equipment and its number, and if possible reference the hire agreement number
  - the date and time when the machine is off hire
  - where the machine can be collected from, or when and how the item will be returned
  - the contact details of the person making the request
- the item should be cleaned
- when the machine is collected, or returned, ensure all the accessories and spare parts are included
- record the condition of the machine, and where possible include

- photographs with the date and time on them
- record the level of fuel in the machine

Summary

Appropriate plant and equipment should be used on the project. This equipment could be hired from within the company or an external company, or it could be purchased for the project. This choice is affected by the availability of the item, and cost comparisons of the different options.

The plant and equipment:
- must not be abused
- must be maintained in good and safe working condition
- should only be operated by authorised, licensed and competent operators
- must be stored in a safe area
- keys should be secured in the project offices
- need daily prestart checks to be completed
- damage should be reported and attended to
- should not be overloaded

Project staff should:
- check the equipment hire conditions
- ensure the correct equipment is ordered
- arrange for the equipment to be safely transported to site and offloaded, and loaded and returned when finished
- ensure that the equipment hours are signed off daily
- check equipment when it arrives on the project
- check equipment before it leaves
- notify the supplier in writing if an item breaks down
- notify the supplier when the item is put off hire
- ensure items are calibrated
- maintain service records

Chapter 8 - Materials

What could be simpler, a drawing is issued to the contractor, the contractor orders materials which are delivered to the project site and the contractor installs them? Or sometimes the client provides the materials and the contractor only installs them. Yes, it should be simple, but in many cases the process goes wrong, resulting in wasted time, delays to the project, additional costs, and upsets in the relationship with the client.

Ordering materials – quantities

When ordering materials it's obvious that the correct quantity should be ordered. Sounds easy, but often projects run short of material which results in problems. For example:

- if the material is imported it may have to be air-freighted at short notice and enormous cost
- the material may not be readily available and take several weeks for the supplier to manufacture the required items
- the material may be no longer manufactured (as may be the case with ceramic tiles), and if the contractor is unable to locate material to match those already installed, they may have to rip these out and replace them with other available material
- ordering small quantities often adds a premium to the procurement and transport
- even if the material is readily available, off-the-shelf from a supplier close to the project, there will be the cost for the contractor's personnel making a special trip to collect it
- sometimes the materials are being installed by a specialist contractor who is now unable to complete their works, which may then cause them to demobilise from site, resulting in additional costs and possible further delays if they can't return immediately due to their other commitments
- shortages disrupt the work since workers employed with the task have to be redeployed to another part of the project, returning only when the correct materials are available
- the procurement of the additional materials and organising their transport absorbs a large amount of additional management time
- the delay caused to the project by the shortage of materials makes the contractor appear disorganised and unprofessional

So why do projects frequently experience shortages of materials?

- Often it's simply caused by the Project Manager, Engineer, or Supervisor incorrectly measuring the quantity from the drawing.
- Sometimes the Designer or Architect has included the quantity on the drawings but they are incorrect and the contractor orders the quantities from this without checking. It's therefore good practice for contractors to check the quantities provided on a drawing, since if it's wrong they could be liable for the error.

- No allowance is made for wastage of the material. Items like ceramic tiles or building blocks will generate wastage due to cutting and breakages. An experienced contractor will know what this wastage will be, which often depends on the actual details of where the product is used. Small areas may require detail cutting, which will generate more waste than large, simple areas. Electrical cables are another example since it may not be possible to use the full length of cable because the project specifications or regulations, will not allow the cable to be spliced, leaving all the short bits of cable to be wasted.
- The incorrect conversion factor is used, which often occurs with earthworks materials when the incorrect factor is used for converting the loose material into compacted material. (This factor depends on the type of material and the amount of compaction required.) When material with unfamiliar properties and characteristics is ordered it pays to seek expert advice regarding what wastage or compaction factors should be allowed.
- No allowance is made to lap the material. This is particularly the case with mesh reinforcing, plastic sheeting or roof sheeting. To minimise wastage due to lapping, or splicing of the materials, it's important to be aware of the standard sizes the material is supplied in. In some cases it may be possible to order materials in different widths and lengths, reducing the number of joints and therefore the amount of lapping.
- Sometimes there is theft on the project, so critical material should always be stored in secure locations.
- Often the material has been incorrectly applied on the project. For instance the product has been applied in thicker layers than those specified, this may happen with paint, asphalt, concrete, joint sealer and adhesives. The Project Manager should monitor the application of specialist products, or products that are used in a large quantity on the project. This will enable timely action to be taken to reduce the thickness and wastage, and if necessary to order more material to make up any shortfall.
- The incorrect quantity could have been delivered. I've seen it happen that a project ran out of a material, the Project Manager contacted the supplier and arranged for additional material to be delivered. Shortly thereafter the remaining material from the original order arrived followed by the additional material requested. This results in excess material remaining at the end of the project and causes wasted effort and cost. It's good practice when an unexpected shortfall occurs, that the reason behind it be investigated before ordering the additional material.

Of course, the opposite can happen too, with a surplus of material left at the end of the project. Almost every project I've been involved with has ended up with materials left over. This is a waste of money because:
- the contractor has purchased material which is not required, although sometimes the material can be returned to the supplier for credit, it's seldom they will refund the full original cost even if the material hasn't been damaged while in storage or transit, and is still in its original packaging

- there are costs of transporting, offloading and storing the surplus materials
- there are the additional costs to transport and dispose of the surplus material (disposing of the material can be a significant cost which may include tip fees, and, with some chemicals, additional hazardous waste disposal fees)

The reasons for having surplus material are:
- similar to the reasons that result in there being a shortage of material
- the client may have changed drawings omitting items after the contractor had already procured them
- the contractor may have accidently omitted the item from the structure
- that it may have been applied in thinner layers than specified
- that it may have been mixed incorrectly

It would be pertinent to investigate the cause of the surplus material since some of the above reasons could have serious consequences for the contractor.

When ordering bulk materials like fuel, cement, concrete aggregates or road materials, nearing the completion of the project it's prudent to carefully plan deliveries, ensuring there is only a minimal amount of unused material left on the project.

Ordering materials – specification

When ordering material the full specification of the material must be provided to the supplier, which may include special welding and painting specifications and colours. If the client has not provided the full specification the contractor should request it, alternatively the contractor may propose a product, in which case the product's specification should be submitted to the client for approval. The fact the client has failed to provide a specification for a product does not always mean the contractor can use their discretion in deciding what product to provide.

It's important that material ordered will meet or exceed the required specification, but at the same time it's a waste of money to provide a more expensive product, of a higher specification than the project requires.

Case study:

One of my projects required a couple of kilometres of pipes and the Project Manager sent the drawings to only one supplier, who quoted to supply the standard product they stocked. The Project Manager placed the order for the pipes without checking the details on the quote. The pipes were delivered and installation had started when we discovered the pipes supplied were of a much higher specification than was required. They cost the project twenty-five thousand dollars more than the pipes of the correct specification.

The lessons from the above are firstly, the Project Manager should have obtained at least three quotes, and then it would have been obvious that this quote was more expensive than other quotes for a product of the correct specifications. Secondly, he should have compared the quote with the tender allowable, and since the quote would have been higher it should have alerted the Project Manager to the problem. Lastly, he should have checked the quote and confirmed the supplier had priced material of the correct specifications, after all, materials of a lower specification could equally have been provided which would have been a bigger problem.

Ordering materials – transport

Before placing an order the Project Manager must plan how the item will be transported to site, be offloaded, moved on site, and be installed. I've heard of cases where large items of equipment have been ordered and manufactured only to find that they cannot be transported to the project. Therefore, certain things may need to be checked before ordering, for example:

- Will the item fit on a regular truck? If the item is too large, a special abnormal load may be required.
- If an abnormal load is required will that load be able to travel to the project? The item may be too wide, high or heavy to travel the required route, and unable to pass under or over bridges.
- Some items may require the installation of additional temporary bracings to ensure the item does not deform or deflect while being transported and handled.
- The item may require special packaging for protection, or to be covered to prevent water damage during transport and installation.

If the item is too large to transport it may have to be manufactured in smaller sections and assembled on the project. Before manufacturing the item in smaller sections:

- obtain the client's approval for the fabrication and transport methods
- check the design since it may require modification to allow for the additional cleats and fixings required to assemble the item on the project
- ensure the trades will be available to assemble and commission the item on the project

Ordering materials – offloading and handling

When ordering materials check:

- Will there be a crane available with sufficient capacity to offload the item and place it in the correct position?
- Will there be access for the crane and delivery trucks? If the item is too large or too heavy, or the crane is unable to be positioned close enough to place the item, it may again be necessary to manufacture the items in smaller parts and assemble these in position.
- An Engineer should check that the lifting points are sufficient for safe lifting of the item. These must be capable of supporting the load and should be positioned to ensure the load is balanced and remains level.
- Will special lifting and spreader beams be required?
- Is the project ready to receive the item and have they planned how to offload the item? I've often had items arrive on site which couldn't be offloaded resulting in the transport being kept for days on site while the team made the necessary arrangements. This was costly, delayed the project and was embarrassing.

Ordering materials – installation

It's important to check that the item can be installed when it arrives.
- If the doors or access points in the structure are too narrow to get the item in, and it can't be installed before the structure is complete, it may have to be manufactured in sections which can be more easily handled.
- If the structure is completed before the item is installed there may no longer be access, or space for the crane (for instance, the roof may have been installed and preclude the use of a crane).
- Surrounding structures may be under construction and prevent a crane being placed close to where the item will be installed. If so either these structures should be delayed, the item must be installed at an earlier opportunity, or be manufactured in smaller pieces.
- How will the item be secured in place? This is particularly relevant for structural steel items fixed in concrete which must be secured in position before the concrete is cast. To assist with this, request the fabricator:
    o provide additional bolt holes, or cleats, to enable the item to be secured to the formwork
    o supply bolts which are already made-up in sets to maintain their positions relative to each other while the concrete is being placed
    o supply a template to secure the bolts in position until the concrete has set

Ordering materials – delivery dates

Often suppliers are not told when a product is required, resulting in it arriving late, or the supplier rushing fabrication, at an additional cost, and it arriving ahead of time.

The ideal situation is to have the item arrive exactly when it's required, because:
- on most projects the client does not pay for material not fixed in position, yet the contractor has to pay the supplier when the item is received, creating cash flow problems
- materials stored on site could be damaged or stolen before they are installed
- materials, like reinforcing steel, may corrode if exposed for a lengthy period in a corrosive environment (like near the ocean or on certain industrial facilities)
- the area may not be ready for the material and it may have to be offloaded far from where it's required, which may result in double handling
- for some items of equipment the warranty period starts from the date it's dispatched to the project, while the required warranty period will normally only start from when the project is completed and handed over, which means if the product develops a fault outside the manufacturer's warranty period, but within the project's warranty period, the contractor will be liable for the cost of the repairs

However, there is a risk when materials are ordered to arrive just in time there may be delays due to:
- problems with the fabrication

- strikes or industrial problems at the suppliers
- suitable transport not being available when the item is ready
- bad weather which impacts the manufacturing process or transport routes
- the item being fabricated incorrectly or not meeting the required specifications, and no time to have a replacement supplied

In addition the project may be running ahead of schedule resulting in the item being required sooner than was originally anticipated.

Weigh up the project requirements carefully before specifying the delivery date. In general, I prefer for long-lead items, manufactured specifically for the project, to be delivered as soon as possible.

Another way of dealing with the risk of late delivery is to get the supplier to manufacture the items and store them at the place of manufacture. Alternatively, when space on the project is limited, the contractor could establish an off-site yard to store long-lead specialist items – (but of course, this comes with the additional costs to rent the facilities, arrange security, and have loading and off-loading equipment available).

An important aspect of specifying the delivery date is also to state the sequence in which items should be delivered. I've had a project order reinforcing for a structure and the first deliveries have been the walls or columns, with the foundations delivered later. Equally it often happens that pipes are delivered first and the couplings and fittings for the pipes come later, or structural steel suppliers send out the heavy steel first while the bracing and smaller items follow. All leaving material on site which cannot be used because not all the components are available, or the components required first have not been delivered.

Ordering materials – quality procedures

The supplier must be provided with the full quality procedures they should comply with, this may include particular tests or procedures that must be carried out. These tests may also have to be witnessed by the contractor, or even the client.

Case study:
I had a major supplier almost complete their contract when the client asked for welding maps, welder trade certificates, and the weather records for the days when items were painted. Needless to say, since the fabricator didn't take note of these items at the time of manufacture, it was impossible to produce some of this information.

Quotes and tenders

Before an item is ordered the project team should obtain quotations for the item – more than one and preferably at least three. Often Project Managers are in a hurry to have the material delivered and place the order with the most convenient supplier without first obtaining other quotes.

A good starting point in this process is to obtain quotes from the suppliers who priced the project at tender stage, but bear in mind these may not necessarily be the cheapest. Investigate if there are any local suppliers, since a supplier near the site may save on the item's transport costs. Also ask other Project Managers which suppliers they would recommend.

When requesting a quote include the following:
- date of the request
- company name
- project name
- name and contact details of the person requesting the quotation
- details of the product required (include specifications, sizes, paint finishes, and so on)
- when the product will be required
- any special requirements
- quality procedures and paperwork required
- design and drawing information
- delivery details, (including the address of the project if the supplier is quoting to transport the product)
- when and where the quotation must be delivered, and to whom it should be addressed

With major components it may be necessary to send out a formal tender enquiry. The documentation would take the form of a full contract document and there would normally be a formal tender closing date and tender adjudication process.

Is the cheapest really the cheapest? – (adjudicating quotations)

When adjudicating the various quotes:
- ensure the quotes are for the same (or similar) product
- ensure the item will meet the standards and specifications required
- be sure all items have been included in the quote
- understand the cost of transport
- check for hidden costs, like insurances and design, which may have been excluded
- ensure the project is able to comply with the supplier's conditions of quotation
- check the supplier can meet the delivery date
- it may be pertinent to check on the supplier's previous experience and even carry out reference checks with their past customers, because I have on occasion, placed an order with the cheapest supplier only to find half way through the project the supplier had a problem (which included insufficient cash flow, lack of manufacturing capacity, being unable to secure material from their suppliers, lack of skilled personnel, lack of experience, or a poor understanding of the project requirements)

Compare the quotes with the tender allowable to ensure they are within budget. If the quote is different from the tender allowable (either lower or higher), it's good practice to analyse and understand why there is a difference. Sometimes the tender is incorrect, and the allowable is too low, in which case the project will lose money on the item and the budget must be adjusted to take this loss into account. If this is the case the estimating department should be advised of the problem so they avoid repeating the error on another project.

Purchase orders

Suppliers must be issued with written purchase orders which should:
- be clear and unambiguous
- have the project name
- have the date of the order
- include an order number
- have the suppliers' name and contact details
- include a complete description of the product
- reference any standards, specifications and drawings the product must comply with
- have any specific manufacturing instructions and details
- specify the delivery date
- include the arrangements to transport the item (if the supplier is providing the transport include the full delivery address, instructions and possibly even a map)
- have the product price, specifying the unit of measurement and what is included in the price
- specify the terms of payment, as well as any trade discounts
- include the address where the supplier should submit their invoice
- specify the warranties and spare parts required
- include the name and contact details of the person issuing the order
- be signed by an authorised person (this may depend on the value of the order, since often people are only allowed to sign orders up to a particular value, and may have to ask more senior management to sign orders with a greater value)
- be acknowledged by the supplier so there is a record that the supplier has received and has accepted the order

Copies of the orders should be kept on file at the project, and be referred to so the project can track the fabrication process, ensuring the material is delivered on schedule.

Samples

Sometimes it's good practice to request a sample of the product from the supplier so it can be used to confirm the quality, obtain the client's approval, and check the colour if this is an issue. In addition, by retaining the sample it can be used to monitor the quality and compliance of the items when they are delivered.

Batches

For large quantities of product the supplier may have to manufacture a number of different batches. These batches may vary slightly in quality and colour. A colour variance can be a problem for items like tiles and paint.

Each batch should have their own separate quality checks and paperwork so any quality deviations can be tracked to the relevant batch.

Chapter 8 - Materials

Procuring from a foreign supplier

When materials are procured from a foreign supplier the Project Manager must ensure:
- the items will comply with the project's standards and specifications (other countries use different standards)
- the items are compatible with local products
- spare parts will be readily available
- warranties and guarantees will be valid
- there are no additional costs for the importation of the items, such as additional taxes and import duties
- the transport costs are factored into the overall cost
- the cost for staff to visit the factory, to ensure the quality standards are met, has been taken into account

Consideration also may have to be given to the stability of the foreign country, since disruption to the manufacturing process (due to strikes, unrest and war) will probably not be grounds for an extension of time claim.

Source of materials – environmental, health, safety and legal considerations

When materials are ordered consideration must be given to the source of the materials and the manufacturing process. Products must be manufactured using materials that will not cause a safety or environmental hazard when being worked with, during the life of the facility, or even when the facility is finally demolished, since these could result in costly legal problems for the contractor.

When procuring materials excavated or mined from quarries or borrow pits, it's the contractor's responsibility to ensure that the process of extracting these materials has the required permits and authorisations in place (even if the operations are carried out by a reputable supplier). If they aren't it could be embarrassing for the contractor, result in fines, and almost certainly lead to an interruption of the supply, which will delay the project.

The contractor should also ensure that materials are not procured from manufactures who are performing an illegal operation, or suppliers that are operating in an unsafe environment or not paying or treating their workers according to the required laws and standards. Not only can this be awkward for the contractor should it be exposed in the media, but it could result in the manufacturer being closed down, thereby affecting the supply of the materials. In addition, the contractor has an ethical duty to ensure the project is not contributing and supporting unlawful practices.

It's advisable that major manufacturers and suppliers are visited before the order is placed to ascertain that the facilities are adequate to produce the quantity of product, to the required standard and schedule, and that the facilities are not obviously violating any rules and standards.

Impartial tender process

The Project Manager should ensure project staff follow a fair tender process, and do not favour a particular supplier because they have been unduly influenced by them, have been given a gift, or invited to sporting events, promotional visits, or functions by them.

Care should be taken to ensure orders are not placed with suppliers who are relatives or friends, unless it can be proven they can deliver the product on schedule, to the required quality standards, and at a lower cost than their competitors.

Alternative materials

It's often worth investigating alternative types of materials because these may:
- be cheaper
- be more readily available
- be easier to install
- be a superior quality
- create less wastage

Case study:

One of our projects used a clay brick from a supplier in the vicinity. The brick, however, was a poor quality and broke easily, resulting in high wastage, and their size was inconsistent, which made them difficult to work with. After investigations we found a product which did not break so easily and was uniform in size. Although the alternative was more expensive, the overall cost was less, because it had lower wastage and resulted in better productivity.

Usually before an alternative product can be used, the client has to give their approval for its use. No client will accept a less favourable product unless there are going to be cost savings for them. It's therefore essential that you provide the client with the product data sheets for the alternative, and compare the performance with the original product.

Expediting and managing material deliveries

Once the material has been ordered the process of expediting the deliveries has to be co-ordinated since it often happens that the project team orders the material then forgets about it until it's required, which is often too late. There is then usually a frantic call around trying to find out where the item is (sometimes the item may, in fact, have been delivered but can't be located since the appropriate people aren't aware of the delivery).

The material orders should be put in a filing system, and the delivery dates should be monitored. For critical items a responsible staff member should be communicating with the supplier regularly, and at least the week before the item is due, to confirm the delivery is on schedule.

Manufacturing facilities should be visited frequently to ensure the fabrication is on schedule. I've had suppliers tell me the item was being fabricated, only to find the process had barely started. If you're unable to visit the facility insist on regular reports from the fabricator, with photographs (it's difficult to make photographs lie) showing the progress.

Transport of materials to the project

It could be better if the supplier transports the item to site because:

- they will be responsible for the item up to the point of delivery and accept the risks for the transport (for expensive items that may get damaged in transit this may be better for the contractor)
- some equipment may require specialist transport which the supplier has, or has access to a suitable specialist for transporting their goods
- if the contractor supplies the transport, the supplier may delay loading these vehicles until they have loaded and despatched their own vehicles (this is often the case when collecting gravel or stone from a quarry, which adds to the cost of the transport since fewer loads will be transported due to waiting time)

Carefully consider all options before deciding who will transport the materials. Generally, though, the project may be able to save costs by arranging their own transport since the supplier often uses an external transport company and will add their mark-up to these costs.

Make sure valuable items are covered by insurance while they are in transit in case they are damaged or stolen.

When importing materials, or when items are transported by ship, a specialist transport company may have to be engaged to take care of the shipping and customs formalities.

Ensure that transport companies are aware of any restrictions in accessing the project such as:
- drivers or trucks requiring access permits
- trucks requiring specialist safety equipment
- drivers needing personal protective gear
- correct licensing and paperwork
- restrictions on delivery times (for instance, construction sites in the city may not be able to accept deliveries during peak-traffic hours, while other projects may not accept deliveries after-hours)
- any restrictions on the access road, for example width, height, length or load limitations
- vehicles having to be escorted on the roads approaching the site
- vehicles may be unable to use the roads in certain weather conditions

Very obvious, but something regularly overlooked, is to make sure the transport company has directions to the project, and contact details of a project staff member. Sometimes delivery trucks get lost resulting in loads going to the wrong site, or trucks getting to site later than planned and disrupting the offloading schedule.

Case study:

I heard of one example when a delivery vehicle travelling to a remote site, took a wrong turning. The driver realised after a while he was on the wrong road and tried to turn the vehicle around, but since the vehicle was long he had to go off the road to turn, and consequently got bogged in the sand. This happened on a remote road which was seldom used and had no mobile telephone reception so the driver had little choice but to leave the vehicle and walk back to the nearest main road where he hoped to find help. Unfortunately the temperatures were in excess of 40⁰ C in the shade – not that there was much shade in the area. The driver was not prepared for such an event, didn't have much water or suitable clothing. Since he had wandered off the designated route he was only found several days later – dead from dehydration.

Since transport can be a major component of the cost of the materials it's usually advantageous to look at alternative transport companies or means of transport. In many cases, we were one of the first contractors to arrive on a project site and one of the first to demobilise so were often able to demobilise our equipment from the project using the transport that was bringing materials and equipment to the project for the follow-on contractors. This saved us money because we weren't paying for an empty truck to travel to the project to remove our equipment, and the transport company was happy because they got paid twice for their return leg.

Case study:
One of our projects required clay bricks from a manufacturer more than two hundred kilometres away. The cost of the transport would have equalled the cost of the brick. Adjacent to our project was a sugar refinery that had a fleet of trucks hauling sugar along a route that passed the brick supplier. These trucks returned empty from their deliveries, so we were able to strike a deal with the company to collect our bricks on their return journey at a cost of less than half of what we would have paid for other transport. This was a win for our project, as well as for the transport company.

Of course, like everything, cheapest is not always best – but often it is. Still you need to ensure the materials will not be damaged in transit, and that the transport company is reliable, and has trucks that won't break down en-route. We've also had transport companies whose trucks arrived on site leaking oil, which was obviously an environmental issue on the project.

Alternate modes of transport should also be considered, in the past I've used both rail and ship. It's often easy to load a container and send it by rail, although consideration has to be given as to how the project will load and offload the container, since containers can be difficult to unpack, which often requires additional labour.

Transport of materials and equipment cross-border

Transporting of equipment and material across international borders should be done by a specialist that understands the paperwork required, as well as the taxes and duties that may have to be paid. It will assist if the company has an agent at the border post that can expedite the process. At some border posts in Africa it can take up to a week for a truck to get through, and if the item has to cross several borders en-route to the project it could take several weeks for the transport to cover a couple of thousand kilometres. The process will be delayed even further if the paperwork has not been completed correctly, it may even result in the truck not being allowed through the border post and having to return to their point of departure.

It's important, that equipment supplied to construct a project in a foreign country is able to be returned, on completion of the project, to their home country without incurring any additional taxes and duties. I have heard of cases of a company not being able to remove their heavy earthmoving equipment from a foreign country without paying very expensive duties. Needless to say all the export and import paperwork must be done correctly, and the company freighting the equipment has to understand which equipment will be permanently included in the project and which will only be used for the construction phase of the project, then be returned to their country of origin.

For some projects the client may have negotiated with the government that materials and equipment incorporated into the project can be imported duty-free. This requires either there are arrangements in place at the border crossings, that with the correct paperwork, allows the material through without any duties being paid, or a system whereby paperwork is submitted enabling the contractor to be refunded the taxes and duties paid. All import and export documentation must be filed correctly and maintained on the project.

Receiving and storing materials

Often materials arrive on a project site when no one is expecting them and there are no plans in place for where the material should be offloaded or how it will be unloaded. Sometimes these materials are delivered after-hours when there is no one to receive them. Consequently this often results in the item being offloaded in the incorrect location and I've even had cases of projects accepting materials that were not meant for them. This results in double handling since the materials have to be relocated, and the cost of this can be significant if a special crane or truck is required to move them.

When deliveries of material are expected:
- ensure staff know where the materials should be placed
- that the area where they will be offloaded is suitable, for example:
    - it may need to be levelled (some large items of equipment or buildings can be damaged if they are not stored in a level position)
    - ensure the ground is firm enough to support both the item of equipment and the crane offloading it
    - the area should be well-drained to enable the area to be accessed even after rain
    - check the delivery vehicle can access the area, and modify the entrance roads if required
- (if required) have packers or bearers available to place under the item so it doesn't sit directly on the ground, and to enable lifting equipment to be removed from under the load
- ensure the required special slings and spreader beams are available for heavy loads, and that lifting studies have been approved
- check weather conditions, since rain may make parts of the site impassable or high winds prevent cranes from operating, and if necessary delay the transport of large items
- if the item of equipment is being placed directly in its final position check the area is ready, for instance concrete structures should have all the quality checks completed to verify the concrete has reached the required strength, the structure has been constructed in the correct position and height, and all holding down bolts are in their correct positions and the

right size (there will be nothing worse than having a major item of equipment suspended from a crane, when you discover the item doesn't fit due to an error with the concrete construction)

The person receiving and checking the material being delivered should be both competent and reliable. This person should:
- verify the item delivered is the same item as written on the consignment note
- verify the quantity and note discrepancies
- know what to check for, and if material is defective or unsuitable should not accept it or note defects on the consignment note and immediately confirm them in writing to the supplier with photographs if possible
- ensure the item is offloaded in the correct place
- ensure the item is offloaded and handled correctly, using the correct equipment, thereby avoiding accidents and damage to the item
- ensure the item is offloaded and stored in such a way that it will not fall over
- when the item is placed, either in its temporary or permanent position, protect it so it cannot be damaged by passing vehicles, other machinery or activities in the area, because I've seen long lead, critical and expensive items get damaged resulting in additional costs and delays

Of course storage doesn't just affect large bulky items, and all storage areas should be planned and set up in a way that:
- materials can be received and stored in a logical and safe manner
- delivery trucks are able to enter and exit the area
- the material is readily accessible when required
- flammable and dangerous liquids are stored separately
- sees all liquids stored in areas with bunds that will retain liquid that's spilled
- ensures sensitive materials such as certain PVC items, are kept out of the direct sun (I've seen PVC pipes stored in the open, exposed to sunlight, become damaged and condemned by the client)
- protects items from weather damage, rain or dust, with either the supplier providing it in sealed containers or shrink wrapped in plastic, in which case these coverings should be checked to see they aren't damaged when the equipment is handled, or alternatively the project may have to ensure there are sufficient weather-proof structures in which the material can be stored
- allowing certain products to be stored within a required temperature range
- older material be used before the newer material, this is particularly the case with cement and products with a limited shelf life or expiry date
- prevents heavy loads from being stacked or stored on scaffold or concrete suspended slabs unless these structures are capable of supporting the additional load

Reporting of defective materials

It's important that defective materials are reported and not incorporated into the works. Steps must be implemented to rectify the defects and prevent them from reoccurring.

(Refer to Chapter 9 for more information)

Contamination of materials

Contamination can be a major problem with concrete and road building materials, so they need to be stored in such a way that they will not come in contact with other materials or be contaminated by the material on which they are placed. Areas should be well-drained with bunds to prevent stormwater running through the area, washing in local soil and carrying out the material. I've had occasions in the past, after rains, that the loader loading the material has picked up mud on its tyres and tracked it onto the concrete aggregate causing contamination.

It's not just necessary to prevent contamination in the project storage areas, you should ensure that the supplier is storing the material correctly to avoid contamination in their facilities.

Contamination can also occur in the transport process, and trucks used to transport different materials should be cleaned out properly before loading a different product.

Material handling

Material handling can play a major role on some projects often resulting in a bottleneck if it's not planned correctly. The project may have the right numbers of personnel and equipment, but if the materials can't get to where they are required, then personnel and equipment will be unproductive and the project will be delayed.

Material management is not only a planning issue, but can also be one of diplomacy, since the Project Manager and Supervisors must juggle the needs of the different teams on the project. Often the various teams and subcontractors can become irate and frustrated due to delays in receiving their materials.

Material handling can be a particular problem on high-rise buildings where items and personnel, including those belonging to subcontractors, must be delivered to elevated locations.

To help handle materials efficiently:
- ensure there are sufficient people and the right type of equipment to handle them
- make sure material transport, loading and unloading is organised and managed properly
- look at ways to assist with loading and offloading, such as getting the supplier to pack and secure the items onto pallets (there may be an additional cost, but it will save time and make handling easier once the item arrives on site)
- it may be necessary for deliveries to take place after-hours when cranes will be available
- offload material close to where it is required
- ensure storage areas are well-planned, organised, neat, with a traffic

management system and good access roads

Supplier's shop drawings

Some fabricated items may require shop drawings to be produced before manufacture can start. These drawings need to be approved by the project team, possibly even the client, and checked to ensure they comply with the project requirements before fabrication begins.

(Refer to Chapter 4)

Materials supplied by the client

Materials supplied by the client should be treated as if they were materials the contractor had ordered. Many times the client is only supplying the item and has given no thought as to how the item will be transported, lifted and fixed in position, which will all become the contractor's problem. I would suggest that well before delivery, the contractor requests drawings of the item to evaluate the item and check:

- its size and weight so as to ensure it can be moved, lifted, placed and secured in the required location, using the available equipment and access, if it can't, it may be necessary to request the client to supply the item in smaller sections
- that the client has allowed adequate lifting points
- if the item requires additional bracing or packaging to prevent it being damaged during installation

Like any other material, when taking delivery of the item supplied by the client:
- check the item for damage
- offload the item and store it correctly
- take the necessary precautions to protect it, because once the contractor has taken delivery it will become their responsibility
- check all the parts are included with the delivery
- check the item to ensure its fabricated to the correct dimensions and tolerances, even if the item is 'client supplied' the onus will be on the contractor to use their best endeavours to ensure it complies with the drawings and project specifications

On site measurements and templates

Some items may have to fit exact openings or sizes on a project, and fabrication of these items may have to wait until the openings are completed so actual measurements can be taken. If there is insufficient time on the project schedule to allow for this templates can be made up to enable the site works to be done according to the exact size as the fabricated unit. Templates can also be useful when constructing multiple units of the same size (for example, window openings or bolt sets) and even to ensure rooms in a hotel or apartment block are built to the exact same size.

When asking for quotations it may be necessary to request the supplier provide a template which can be used on the project for construction.

Chapter 8 - Materials

Summary

When ordering material:
- consider if there are cheaper alternative materials and whether it could be worth the cost savings to use these materials
- calculate the quantity of materials required, allowing for wastage, and using the correct bulking and compaction factors, and application instructions
- request at least three quotes
- compare the different quotes ensuring all costs have been considered, including those for transport
- ensure the supplier has quoted for the correct product and specifications
- check the capabilities of the supplier to ensure they can produce the product to the required quality standards in accordance with the project schedule and that they have all the correct permits and registrations in place
- ensure the item can be transported to the site, offloaded, stored, handled and installed
- have measures in place to protect it in transit to avoid damage
- ensure its handled and stored correctly to avoid damage
- have the manufacturer prepare the required shop drawings which should then be checked by the contractor
- give suppliers orders which clearly outline the cost, specifications, quantity and delivery schedule
- contractors must manage the manufacturing and supply process to ensure quality standards and procedures are followed and there are no delays with the delivery

Chapter 9 - Quality Control

Unfortunately quality control often just turns into a paper exercise, and is a task the Project Manager leaves to the Supervisors, or on bigger projects Quality Engineers or Quality Managers. However, it's the Project Manager's responsibility to ensure that quality control is treated seriously, is not only about paperwork, and that people are delegated with specific responsibilities to deliver the correct quality, understanding what to look for and what the required quality standards are.

All the quality paperwork in the world, with all their signatures, will not turn a poor quality product into a good quality product. However the paperwork trail is important in ensuring that proper quality procedures have been implemented and followed.

Quality is about:
- delivering to the client a project that meets and exceeds their standards and specifications
- the project meeting the local bylaws and codes
- meeting the code and specification requirements of the state or country (except if the client has particular exemptions allowing deviations from these codes and requirements)
- meeting the contractor's own standards

These requirements extend to the installation's:
- functionality
- durability
- suitability
- finished aesthetics

To meet these quality requirements the contractor has to ensure that the materials used, and the equipment permanently installed as part of the project, all meet these requirements.

I'm frequently amazed at how many workers have no pride in the quality of the work they produce. Many homes, hotels, apartments and shopping complexes I've visited show signs of poor quality. I regularly see examples of poor tiling, walls that are built out of square and doors not fitted correctly.

Case study:

An important lesson to me came through a Foreman who was probably one of the best I have worked with. Our client tasked us with placing concrete inside a concrete silo built by another contractor. The concrete had to slope steeply to an opening in the centre of the floor. I'd said to the Foreman that this was a rushed job and the client was only concerned with the structural integrity, not the aesthetics of the work, so he shouldn't waste time with finishing the concrete too neatly. The Foreman was horrified at this statement, and told me in no uncertain terms that as long as he was building something it would be to the best quality possible, and finished off correctly. He would not take short cuts. Even if nobody would see this concrete it would be to the same quality as any other concrete he placed. There was only one way to do a project, he'd said, and that was to do it right.

And, of course, he was absolutely right. This is the way we should all view our work and perform our tasks – with pride!

Case study:
Quality issues are not just cosmetic. One of my projects was to construct a reinforced concrete silo thirty-five metres high and eighteen metres in diameter with external walls three hundred millimetres thick. We constructed the silo using slip-form shutters which involved a continuous concrete pour, twenty-four-hours a day, for a period of twelve days. The formwork to the walls was slowly jacked upwards and fresh concrete was poured into the forms at the top, and as the concrete cured at the bottom of the formwork, the formwork was moved up. We employed a specialist slip-form subcontractor to carry out this work.

Normally, the rate of movement of the wall formwork is between two and a half metres and three and a half metres a day, with reinforcing fixed ahead of the forms as the formwork moves upwards. For some reason we elected to do the reinforcing fixing ourselves using two reinforcing teams, one for the day, and one for the night shift, each led by a reinforcing Leading-hand.

The slip-forming contractor had a full-time Supervisor, we had a dedicated Engineer and the managing contractor had a permanent representative on site, and at night there was a night shift replacement Supervisor, Engineer and managing contractor representative.

There were no reports of problems during the slide and after twelve days we reached the top of the silo and dismantled the formwork. All the quality paperwork was signed off correctly.

A few days later the client noticed several tons of reinforcing lying on the ground which was obviously meant for the silo. Had this reinforcing been omitted from the walls in error? We began a full investigation, firstly checking if the bending schedule was correct or if it had possibly detailed more reinforcing than required, which wasn't the case. We then questioned the supplier to check if they had mistakenly delivered extra reinforcing, which again wasn't the case. The only answer appeared to be that the reinforcing had been left out of the walls, so we started to check them, firstly with reinforcing detectors, and finally as the enormity of the problem hit us we actually cut into the concrete walls to expose the reinforcing.

Everywhere we checked we found the reinforcing was fixed incorrectly and the horizontal bars were placed further apart than they should have been. In fact the problem was so severe that we had installed on average about 25% less horizontal reinforcing than we should have.

One of the many frightening aspects of this mistake was that the reinforcing seemed to be spaced almost randomly, it was not as though the incorrect spacing had been used at the bottom and then replicated for the rest of the silo. Secondly, every shift seemed to repeat this arbitrary fixing of the reinforcing. Despite having a slip-forming Supervisor, a reinforcing Leading-hand, our Engineer and the client's Supervisor on each day shift, and the same again for the night shift, which meant a total of at least eight persons who should have ensured the correct quality, we had a major quality problem. One of the first things anyone responsible for quality should have checked was the reinforcing spacing, and yet in the course of twenty-four shifts *nobody picked up this fundamental and obvious error. Everyone had failed to carry out this simple routine quality check.*

This was an error of significant proportions. After the Design Engineers checked their calculations they found that the silo with the reduced reinforcing was not suitable, and we would have to take remedial action to rectify the problem. Various options were considered, one was to install additional strapping to the outside, which would have been unsightly, and then there was the worst case option of demolishing the silo and rebuilding it.

We ended up slip-forming an additional outer wall to the silo which cost us nearly a hundred thousand dollars, resulting in a significant reduction in our profit, and, of course, our reputation was damaged.

So what had caused the failure in the quality assurance process? Clearly all of those involved didn't do their job properly. The lesson here is that you cannot automatically assume that people will follow the correct quality procedures – no matter how basic these may be.

In this case, the problem was discovered while we were constructing the silo, but if it hadn't, the mistake could have gone undetected for many years. The silo wouldn't have fallen down, but it would almost certainly have started to crack badly, then we would have been called back to investigate, and when the shortfall of reinforcement was discovered, we would have had to rectify the problem, which would have been even more costly by then.

Sometimes, when the problem is not discovered during construction, the result can be catastrophic; structures may collapse, disrupting the client's activities, resulting in a loss of earnings, and in many cases causing injury and even loss of life. The Project Manager will be held responsible if it's found the structure was poorly built, compromising the structural integrity. Even if they were not directly responsible for the quality inspections, and had delegated these to another member of staff, the Project Manager will almost certainly be fined or even imprisoned. For this reason I recommend that Project Managers take the quality of the work on their project seriously since a collapse of a structure, due to a mistake in the building process, will damage both their, and the company's reputation.

The quality process on a construction site is more complex than the process in a factory where there are the same people and machines, all producing the same product in a controlled environment, to the same specifications, as was produced yesterday, probably last month, and even last year. The construction process is far more difficult than this. Every project is different with different specifications, materials and equipment. The people employed on construction projects change from one project to another, and there is seldom continuity of management or craftsmen. The project conditions vary from site to site, and even from day to day, with extreme heat on one day, followed by wind and dust, and maybe rain the next day. However, the construction team is expected to overcome these difficulties and produce a consistently good quality product. This presents major challenges to the Project Manager, and the only way to ensure a quality product is produced is to take the lead in quality management, and ensure that the correct quality systems are implemented on the project.

Chapter 9 - Quality Control

Responsibility

The Project Manager should provide guidance as to the quality requirements and expectations on the project. Whenever they travel around the project site they should be looking with a critical eye at both the completed work, and the work under construction, to ensure it conforms to the project's quality requirements.

A top-to-bottom commitment is required for quality. A project will not achieve the required quality standards if individual workers are not committed to producing a quality product. At the same time, even if the workers are committed and well-trained, the project will not achieve the desired quality if the Supervisors, Quality Engineers and the Project Manager aren't committed to producing a quality product.

All workers and staff must take responsibility for producing the best quality product possible. Often on projects I see people blaming their tools, materials, equipment, the schedule, the Supervisor, or fellow workers for a poor quality product. Each individual must understand that they are personally responsible for the quality of the product they produce, and they shouldn't play the 'blame game'.

In saying this, management must also be mindful of how they influence the quality of the workmanship on the project, and what they can do to improve the quality culture, like sending craftsmen or Supervisors on appropriate training courses. They should also continually look at the construction process with a critical eye, to see if changes would improve the quality of the end product. Maybe the materials or equipment are genuinely substandard making it difficult for the workers to achieve the desired quality.

On many projects I see completed quality work being damaged by follow-on trades and contractors. All workers on the project should not only take pride in their work, but also in the work done by the others, and they should respect each other's work. Taking a little extra care while working around completed work, and protecting it, will ensure it's not damaged.

There should be a clear delegation of the ultimate responsibility for the quality of a task. Often I've had Engineers, or Supervisors, tell me that they weren't responsible for the poor quality of their section of works and that it was the Quality Manager who was responsible. Obviously this is rubbish! Each Supervisor is responsible for everything within their section of works, including the quality of work and the materials used. Every Engineer should be responsible for the quality of their section of the works. The Quality Engineer, or Quality Manager, is appointed to assist the Supervisors and Engineers to monitor and record the quality, and to ensure the required quality systems are implemented and followed.

Poor quality should not be accepted, and Project Managers must not pass by substandard work without taking action. This action may be to simply chide the responsible party for minor defects, but with serious breaches of quality, consideration should be given to implementing disciplinary procedures against the parties responsible. Consideration should also be given to demolishing the substandard work, although many Project Managers loathe doing this, I've learned over the years that it's often the right course to follow. Supervisors are extremely embarrassed when they have to demolish their substandard work, and usually ensure that work they do after this conforms to the required standards. In addition, substandard work that is repaired often shows the patch for the life of the facility, and creates a weak point in the structure which may later require remedial work.

Quality plans

The quality plan must be drawn up at the start of the project, then as the project progresses, it may have to be modified to include new processes previously overlooked, or unforseen, at the start of the project.

The project quality plan is a document that sets out the quality requirements for the different project processes. Therefore it should firstly list all the processes that will take place on the project. These could include setting out, excavation, pouring concrete, backfilling, erecting steel, bricklaying, and so on. I normally use the project schedule as a guide to what processes are involved in the project.

The quality plan should be no more than a few pages long, although there's much back-up documentation which has to be read in conjunction with it, and should be referred to. This documentation includes the project and other relevant specifications, inspection forms, and checklists which have to be completed to ensure the processes are completed and conform to the required specifications.

The quality plan should refer to the importance of the check, and who is responsible to carry out the check, and whether the work can proceed without the client having to approve the inspection. When I prepare a quality plan I try and put as much of the control for the quality in the contractor's hands. The more control (or hold points), there are in the client's hands the more chance there is of delay should the client not be available to carry out the inspection.

Once the plan is completed it is normally submitted to the client for comment and approval, and they may make changes to take more control of the processes.

Quality documentation

The quality system must be simple and easy to implement so that it doesn't hinder the production process. It must be easy enough for the construction staff to understand the requirements and correctly complete the documentation, and it shouldn't be a time-consuming process. Therefore the Project Manager and the construction team should work together to produce meaningful, simple checklists and quality documentation, relevant to the tasks on the project, that will enable the process to be implemented, recorded and monitored.

The quality documentation should not be about creating rules or fear, nor should it be treated as such by the staff implementing the systems. The documentation is about improving processes and recording them.

It must also comply with the project quality plan and project specifications. It's pointless carrying out an inspection in accordance with a document if the inspection doesn't comply with the specifications. I've often seen documentation and templates used on another project simply rolled over onto the new project, with no checking done to ensure the documentation complies with the requirements of the new project.

Documentation must also meet the requirements and expectations of the client and their representatives. I would normally at the start of the project agree with the client what form the quality documentation should take. Agreement must be reached as to what sections of work will be grouped together in a file, what information will be included in the file or dossier, and I would normally submit an index for a typical file and get them to agree to it. Many of the client Quality Managers can be pedantic as to how they want the quality information presented, so it's always good to agree all of this up front to prevent the quality documentation from being redone at a later date to conform to the requirements of the particular individual. Having said this, however, sometimes the client has expectations in excess of the contract requirements. Should the client request additional controls, tests or checks over and above what is specified in the contract, or over what the norm would be, then the Project Manager should consider submitting a variation for the additional costs involved.

From the start of the project the construction team must understand what the requirements are to achieve acceptable quality on the project, and the quality documentation required so it can be collected as the project progresses. I've seen many projects where this has not been done, resulting in a mad rush at the end of the project to find the missing paperwork. Often by this late stage they aren't available because the relevant tests and inspections were not done.

The documentation is a record of the process, it's not the inspection. I often see the quality documentation just become a 'tick and flick' operation which a Supervisor or Engineer hurriedly completes in the office, without actually inspecting the work. They must understand that completing a checklist on a piece of paper is not the same as actually conducting a thorough inspection of the product, and the completion of a piece of paper does not absolve them from the responsibility of the inferior quality of the work.

Non-conformance reports

Non-conformance reports are reports that are completed when a process, or task, has not conformed to the required quality standards. This may be as a result of failed tests on the item, that visually and aesthetically it does not conform, or that quality steps in the process were not followed, or aren't traceable.

Most contractors are fearful of issuing non-conformance reports to themselves for non-compliant work. However, the purpose of the report is not to attach blame, or create fear, but is rather about creating awareness of the problem, seeking appropriate corrective action, and ensuring that there's an improvement in the process so the problem does not occur again.

Frequently Supervisors make an error and quickly cover it up with a patch which results in:
- the cost of the error not being tracked
- the error possibly being repeated by the Supervisor, or another Supervisor
- the repair often being inappropriate, which can lead to defects and weaknesses later in the life of the project, which will then incur additional costs to repair

The non-conformance report requires the following information:
- the date the report was compiled
- the location of the problem
- who compiled the report
- the nature of the problem (refer to the relevant specifications and include,

if applicable, photographs, test results and reports)
- the cause of the problem
- how the problem should be rectified
- the estimated cost to repair the problem
- what action should be taken to prevent the problem from recurring

A register should be maintained for the non-conformance reports to ensure they are actioned, closed out correctly, and that the problem is rectified.

Once the non-conformance report has been completed it should be circulated to the relevant project staff to enable the appropriate actions to be implemented. The report is normally issued to the client and the contractor's quality department.

Sometimes the problem can't be easily rectified, and then it's necessary to investigate how it impacts on the facility. Depending on the investigation's outcome the client and their Engineers may accept the item in the non-conforming state.

Errors made on construction projects cause an enormous amount of rework, resulting in unnecessary costs, delays to the project, and compromising the integrity of the structure. By tracking these costs and delays staff will better understand how important it is to ensure the work they do conforms to the required quality standards and specifications, so hopefully in future, more care will be taken to prevent a repeat of these errors.

Quality of materials

Of course quality is not just about doing the work to the correct standard and quality, it's also about ensuring the materials used meet the required standards and specifications as well.

Case study:

Another concrete silo we constructed stored rock ore from a platinum mine. The ore was fed through steel gates into train trucks that passed under the silo. These gates were attached to a metal frame which was cast into the underside of the concrete slab supporting the silo. The silo had been in operation for a few months when we received a call from the client to say it could not be used because the frame was coming loose from the concrete. The client had inspected the frame and found that the welding did not comply with the specifications. We had to urgently engage a specialist contractor to drill and epoxy grout the frame back into the concrete, so the silo could function again, and the mine continue production.

On inspecting the frame I was convinced there was in fact a design fault with the frame (which would have been the client's Engineer's problem), and that it would probably have pulled apart anyway. However, there was also no doubt that there was a fault with the welding, and we therefore repaired the frame at our cost, which was more than 10% of the contract value.

Case study:

On another occasion houses we were constructing were in their finishing stages, with walls painted and carpets on the floor, when we picked up that one of the doors did not fit correctly in its frame and would not be accepted by the client. After trying to refit the door we eventually found the steel door frame was, in fact, out of square, with one side longer than the other. Neither the Supervisor nor any of the following trades picked up that the frame was out of square, or if they did, they didn't report the error.

The only way to repair this problem was to remove the defective frame and replace it with one correctly manufactured. This happened at considerable cost since we had to replaster the wall, and repaint and carpet the room. Of course, the manufacturer only supplied a new frame and took no liability for the costs incurred to replace it.

The important lesson is that quality checks need to be done on all materials supplied to the project to ensure they are the correct dimensions and meet specifications and tolerances.

Suppliers have to be able to produce the required quality. They should have appropriately trained and certified tradesmen, and relevant and certified quality systems in place.

I've had cases where materials have been procured and delivered to a project, only for the client to reject them because they didn't have the required testing or inspections done during fabrication. The rejection of material, or equipment, causes disruption to the construction schedule resulting in additional costs, which at a minimum includes the additional transport costs to supply the replacement product and to remove the incorrect one, as well as the costs of loading and unloading them.

Reporting and tracking defective material

When material is delivered to the project site it should be checked to ensure it conforms to the required standards and specifications. What happens if it doesn't comply? This is where most projects fall down! In many cases nothing is done, and the material is left where it was offloaded. Although in some cases the Supervisor, or person who received the material, may telephone the supplier and tell them to replace the items.

When defective or incorrect material is received:
- where possible it should be sent back to the supplier on the transport that delivered it
- the supplier must be notified in writing (preferably using a non-conformance report), and this document should record:
    o the supplier's name
    o the date
    o the delivery note number
    o the person who received the defective material
    o what the defective item was
    o what the problem was
    o where possible, photographs
- follow up with the supplier to understand why defective material was sent and how this can be prevented in the future
- it should be clearly marked as defective so it's not inadvertently used on the project
- the Contract Administrator and Project Manager should be notified so they can notify the supplier of costs incurred to load and return the materials, or dispose of them, ensuring the costs are tracked and back-charged

Training to achieve better quality

It's important to have craftsmen on the project who are trained and capable of constructing the project to the required standards and specifications. Without them it's difficult to meet the required standards, and it will be necessary to implement a training program for existing project personnel or new recruits.

In addition, supervisory staff should be adequately trained and experienced (preferably with work experience on similar tasks and projects), and have the ability and knowledge to supervise the craftsmen so as to ensure the project is constructed to meet the required standards and specifications. Sometimes these skills are not available, so it will be necessary to train them. The Project Manager and other qualified staff should also mentor the less experienced staff.

It's also necessary that Supervisors receive training on the quality standards and procedures being implemented to record the quality processes. These may vary from project to project and it shouldn't be assumed that the project staff, no matter how experienced or senior they are, will automatically understand the requirements and systems in place for the project.

Subcontractor quality

Subcontractors are an important component on many projects and as such must be aware of the quality requirements and procedure required on the project, all of which should have been included in their tender scope and contract documents. However, just because it's in their contract document, will not mean the subcontractor's personnel have read them and are aware of the requirements. Often it's up to the Project Manager to make sure they are.

Many times I've found that my staff have ignored the quality of subcontractors' work, and accepted an inferior standard. This is obviously unacceptable, and it's important that subcontractors are notified of substandard work as soon as it's noticed. Problems with the subcontractor's work should be documented, preferably using non-conformance reports, so the problem can be tracked, discussed and recorded in the weekly subcontractor's meeting. This is not about a witch hunt, or to attach blame, it's to ensure the subcontractor, and in particular their senior management, is aware of the issues, and takes appropriate action to rectify problems.

Testing and inspection

Testing must:
- be done in accordance with the quality plan, specifications and contract requirements
- be done at the correct frequency
- use the correct type of test
- in most cases be performed by a qualified independent, registered testing authority

- be recorded with the:
  - date of the test
  - location of the test (frequently we've had tests which failed, but we then had difficulty locating the structure the test results pertained to)
  - item tested
  - method of testing
  - actual results compared to the required result
  - name of the company and the person completing the testing
  - relevant other data, like weather conditions
- be performed by a person who has the correct certification (it's good practice to get a copy of this certification and include it in the quality documentation)
- be done with equipment which is calibrated in accordance with the test requirements (copies of these calibration certificates should be filed with the quality documentation)
- be done correctly, since many failures are caused not by the failure of the product being tested, but rather because the test was not done correctly or the test equipment being damaged

It's important that the results:
- are received by the project team as soon as possible
- are reviewed by a suitably qualified member of the construction team who can take the appropriate actions should the results be below the acceptable requirements
- should be distributed to the appropriate people within the project organisation
- must be filed in the correct place within the quality documentation

The receiving, collating and distributing of the test results can be vitally important and must happen as rapidly as possible, because in many instances, the project processes can't continue until the positive test results are received by the relevant Supervisor or client's representative.

Punch lists

Punch lists are an important part of the contractor's quality process and yet they're often left for the client to produce. On many projects the contractor ends up receiving punch lists from the client containing several hundred, or even thousands, of items that require rectification before the client will accept the works. I've seen contractors complete a project and then spend several months, and in some cases a year, just attending to and completing the items. This has cost them several hundred thousand dollars. In addition, the staff called upon to rectify the items quickly become demoralised since there's little worse than rectifying the poor workmanship of others.

Contractors should be in the habit of preparing their own punch lists as soon as possible after a structure is completed. The Supervisor who constructed the structure should then be tasked with attending to these punch lists. Not only does this mean that the punch lists are attended to while work is proceeding elsewhere on the project, but it also means that the Supervisor responsible for the defect repairs it. Since Supervisors hate repairing defective work, this has the added advantage that they are then more likely to take care with future structures to ensure similar defects are avoided. Also if defects are attended to as the project progresses, the completed facilities can be handed over soon after the last of the structures are completed on the project. But supervisors may require a few additional workers in their team to attend to punch lists while they are constructing the other structures.

Once the contractor has attended to their punch list items, the client should be requested to draw up their punch list. I've had projects where the client has drawn up a punch list, and as the contractor completes the items the client, or other members of their team, adds additional items to the list. This is unacceptable and the contractor must ensure the client only produces one punch list for each trade or structure. The contractor should also ensure the client's representative who prepares their list is, in fact, the authorised person responsible for drawing it up.

The Project Manager should ensure the items on the client's list are legitimate defective items, since often clients put items on the list which were damaged by them, their contractors, or are maintenance items that should be repaired at the client's cost. In addition, I often find that clients add items to punch list which are additional to what is specified in the contract. If items are additional works, or maintenance items, the Project Manager should obtain a site instruction to carry out this additional work. When a client is faced with having to issue a variation for these additional items they usually simply remove them from the list.

Recognition

As important as it is to notice a product of poor quality, it's equally important to notice and recognise one of good quality. I've on many occasions noticed concrete structures of a high standard, and have asked to talk to the responsible person or team, who I have then personally thanked for producing the good quality. (I just wish I had done this as often as I should have.) Personally commending someone in front of their fellow workers for their workmanship has often produced better results than when on occasions I've had reason to criticise workers for their poor quality.

Summary
- Quality is about ensuring that the correct quality documentation is collected from the start of the project as proof that the construction procedures and methods have met the project requirements and specifications.
- More important than collecting the documentation are the physical checks, procedures and tests that must be undertaken to verify the structures meet the quality requirements.
- The materials and equipment built into the structure should be checked and comply with the quality standards. Defective material must be managed and tracked.
- Everyone on the project must be aware of the quality requirements for the project, with everyone equally responsible for ensuring the correct standards are met.
- All personnel must take pride in their work and ensure it's of the highest

possible standard.
- The contractor should prepare punch lists for structures as soon as they are completed, and these should be attended to as quickly as possible.
- Appropriate training should be given to personnel to not only ensure they can produce work of a high standard, but, are also familiar with quality procedures.

The question everyone should ask is, 'Would I pay for and accept this quality in my house?' If the answer is no, then the product doesn't meet the required quality standards.

Chapter 10 - Subcontractors

Case study:
On a recent contract we awarded a subcontract, worth almost thirty million dollars, for the supply of prefabricated buildings, and a few months later nearly doubled the order when we added additional buildings to their scope. A formal contract document and schedule was issued for the original contract and in all respects everything appeared in order, but the new scope was awarded via a two line email which made no reference to the original contract conditions or a schedule. Later, again by means of an email, a further building was added which had a poorly defined scope.

We later found that the original contract was flawed and offered us little recourse or protection if the subcontractor did not meet their obligations.

The fabrication of the original scope proceeded at a slower rate than the subcontract schedule, however, because our works were delayed, there was no concern with the slow fabrication, so we didn't formally inform the subcontractor they were behind schedule or that they needed to rectify the situation.

Only informal meetings were held with the subcontractor, no records were kept of these, and most communications were verbal or, at best, informal emails. Many different staff members dealt with the subcontractor resulting in the subcontractor becoming confused and getting mixed messages. Still, the subcontractor also used this to their advantage by, in certain cases, deliberately confusing the messages, and in other cases selecting the messages they wanted to comply with, and playing our staff off against each other.

Before starting fabrication the subcontractor had to submit design drawings to us which we forwarded to the managing contractor for their approval. We, however, failed to check the drawings, assuming erroneously that the managing contractor would do so, resulting in some instances of drawings being approved which differed from other approved drawings, and buildings then being fabricated in accordance with one drawing but not complying with another.

The subcontractor lodged a number of claims for additional work which they maintained was not in accordance with their original scope. We never responded to most of these, meaning we either lost the opportunity to pass their costs onto the client, or we could have lost our right to respond to the claim.

When I joined the project, the subcontractor hadn't yet designed a number of buildings, some of which were due to be supplied the following month. In fact, they later decided these buildings were not in their scope, because they maintained the project scope had insufficient information to construct them. We were forced to procure these from another supplier, and while they were being manufactured had to supply temporary buildings, all of which cost us several hundred thousand dollars, caused delays to the project, and diverted our staff who had to solve the problem from other tasks.

In addition, the subcontractor queried the specification for one building, and a junior Engineer provided them with the incorrect information, which resulted in problems when the building was fabricated.

The quality of the product produced by the subcontractor was generally poor, but, this was never recorded or reported back to the subcontractor, so no action was taken to rectify the faults, nor to prevent them from reoccurring. Consequently the buildings supplied to the project were substandard, resulting in our personnel having to do substantial repairs to them, costing us tens of thousands of dollars, and taking resources away from other tasks which delayed the overall project.

All of these problems cost us in excess of two million dollars.

This subcontractor was contracted to do over fifty million dollars of work which was a large contract that should have been taken and managed seriously, and yet it contained many lesson on how not to run a subcontractor:

- The subcontractor was selected because they were the cheapest and there appeared to be little validation of their previous track record. We later heard that another large contractor had refused to use them again because the subcontractor had previously let them down.
- The subcontractor was appointed using our company's standard subcontract order, which we later found to be flawed. It's important that companies have their standard subcontract orders and documentation reviewed by a legal expert to ensure that it's legally enforceable, with no ambiguities or loopholes.
- Changes and additions to the scope should have been issued to the subcontractor as formal contract amendments referring to the original contract terms, conditions and specifications, and a schedule should have been issued and accepted for each new portion of works.
- The original contract, and the subsequent amendments, should have referred to a specific scope of works which should have been unambiguous and clear.
- Formal meetings, with minutes, should have been held weekly, or biweekly, and at these meetings the subcontractor's progress, quality, variations and problems could have been discussed.
- Quality control should have been enforced from the start of the fabrication, and:
    - the subcontractor should have been aware of the quality procedures and documents required
    - this documentation should have been collected as the project progressed
    - the subcontractor should have had their quality procedures, systems and documentation audited at an early stage of the project
    - a standard inspection sheet should have been set-up for the buildings, with clearly indicated the 'hold' points where our representative were required to inspect work
    - we should have employed dedicated full-time staff, who understood the quality requirements, to check and ensure the buildings met the quality standards
    - the subcontractor should have been notified in writing of the quality problems
- The subcontractor was allowed to fall behind schedule and we didn't send

them a formal letter requesting that they comply with the original schedule. Instead they were allowed to carry out the works in a sequence and time that suited them.
- We failed to respond to the subcontractor's variation and extension of time requests which caused frustration with the subcontractor since they felt we weren't attentive to their demands. It also placed us at commercial risk because we lost the opportunity to submit legitimate claims which were for the client's account to the client.
- The subcontractor should have been called to the project to be shown the problems, and been given the opportunity to rectify the works. Only if they failed to do this, should the works have been corrected by us, and the costs deducted from the subcontractor's account.
- The subcontractor should have been notified monthly of what the costs of the rectification were in the form of quantifiable, recorded costs.
- Since the subcontractor was not formally notified of their poor quality and non-conformance with the agreed schedule, it was difficult to terminate their contract, or remove portions of their scope because of this non-performance.
- We never checked the subcontractor's shop drawings to ensure the units were designed in accordance with the scope and specification. We simply relied on the managing contractor to do this, which was wrong, because even if the managing contractor approved a drawing it was still our responsibility to ensure it was correct.

Types of subcontractors

There are different types of subcontractors:
- 'Material only' suppliers are contracted to supply only a product or item of equipment, which may also include the transport of it to site.
- 'Labour only' subcontractors are paid per hour, or per day, for the workers they provide to the project, and these workers may be provided with supervision. These subcontractors are paid for the physical hours worked, irrespective of the amount of work completed. The challenge is to ensure the subcontractor's workers are productive, since there is little incentive for the subcontractor to manage their people efficiently and to avoid waste or damage of materials supplied to them.
- 'Labour and equipment supply' subcontractors are paid for the workers and equipment they supply, again the risk lies with the contractor. If they work inefficiently, and use their equipment unproductively, it will cost the contractor more.
- 'Installation only' subcontractors are contracted to install materials supplied by others, and they are paid on the quantity of work or tasks they complete. The subcontractor may supply supervision or the contractor may supervise. There is sometimes little incentive for the subcontractor to look after the materials and they may end up wasting them. The subcontractor may provide their own equipment for the work or the contractor may provide it, in which case ensure the equipment is readily available to avoid the subcontractor waiting resulting in them claiming for this time. Ensure

the contract is clear regarding who should load, transport and offload the materials supplied.
- 'Supply and install' subcontractors normally provide their own material, workers and equipment and are typically paid an agreed sum to complete a task. There is little risk to the contractor, providing the contractor provides access on time and the performance of the subcontractor is managed to ensure they meet the project's quality, safety and schedule requirements.
- 'Design, supply and install' subcontractors usually have to design the whole works, a portion of their works, or an item of equipment that forms part of the project. The contractor must ensure that the design complies with the scope and specifications.
- 'Design only' would be professionals engaged to design a portion or all of the work. The contractor must ensure that the design complies with the project specifications.

All of these contracts need to be treated in a different manner and, in general, the contract document used to appoint them will be different.

In some cases the client may have specified (nominated or selected), a subcontractor that the contractor must use to perform a portion of the works. These subcontractors will still require to be managed, and will be paid by the contractor.

Subcontract tenders

Subcontract tender documentation will vary depending on the type and size of the subcontract. But each should be as clear and complete as possible so there is no cause for misunderstandings which could result in quality, schedule and safety problems, or lead to claims and variations. The tender documentation should:
- have a clear scope of works
- detail all the drawings and specifications that apply to the contract (including a drawing schedule with all drawing numbers and revisions)
- detail the terms of the contract, which would include payment terms and conditions, and insurance conditions
- note special project conditions, including specific project labour agreements and wage rates
- have specific requirements the contractor may have, such as particular staffing or equipment requirements
- clearly state what the subcontractor will be supplying and what they will be provided with, this includes amongst other things, services (like power and water), storage and office facilities, cranes, scaffolding, off-loading facilities, security, and insurances, with any charges for using them specified
- include the contract schedule, which highlights the subcontract activities and any discontinuities they should expect in the course of their work

Selecting a subcontractor

Successful management of a subcontractor starts with the selection process, and they shouldn't be appointed solely on the fact that they are the cheapest, their price was used at tender stage, or they're convenient to use (for instance they are already working on the project or in the area).

The first step in the process of selection is to locate suitable subcontractors which may include:
- approaching subcontractors you've worked successfully with in the past
- asking other managers in the business who they recommend
- checking the subcontractors used for the tender pricing
- asking the client or their representative who they could recommend
- contacting the relevant trade bodies and asking for their list of registered members
- asking Project Managers working for other companies if they know of suitable subcontractors – it's also useful to stop at construction projects in the area and ask the Supervisors to recommend subcontractors
- advertising for them if they cannot be found elsewhere

In selecting the subcontractor consider amongst many things the following:
- Are they experienced with the particular services required?
- Can they produce acceptable quality?
- Have they worked on similar projects or for similar clients? This not only relates to the type of work that is involved, but also to the type of project and client. For instance, many subcontractors may be able to deliver a similar project on a commercial building project in the city, but have no experience on working on remote mining or oil and gas projects which have specific requirements and require more onerous safety standards.
- Do they have the resources to carry out the work?
- Do they have the financial means to carry out a project of this size?
- How does the size of the project compare to the size of projects the subcontractor has undertaken in the past?
- How does the size of the project compare to the overall turnover for the company? Projects that are larger than half the annual turn-over of the subcontractor may result in the subcontractor having cash flow problems, or having insufficient experienced staff available for the work.
- What is their track record on delivering similar projects?
- What other work are they currently doing?
- What is their safety record?
- Do they have suitable staff available for the project?

Past performance, however, is not always indicative of how a contractor will perform on a contract and I have, on occasion, had good subcontractors that have performed poorly, due to them being overcommitted on other projects, which meant they had insufficient and poor quality resources for my project.

Before requesting a subcontractor to tender, you should be comfortable that they have the capacity and capability to deliver the works they are quoting on. This can be done by:
- requesting potential subcontractors to complete a prequalification process which may require them to provide:
  - their safety statistics
  - their financial results
  - a list of similar projects they have completed
  - their number of employees

- - numbers and types of equipment they own
  - current projects they are working on and when they expect these to be completed
- talking to managers in your business who may have worked with the subcontractor previously, or who may have insight into their current situation
- contacting the subcontractor to ascertain if they are interested in tendering for the works, and requesting references
- checking the subcontractor's website and, if these refer to their previous projects, contacting the clients mentioned to confirm they were happy with the subcontractor's performance

The more that can be discovered about the subcontractor at this stage, the better. Not only can the capabilities of the subcontractor be researched, but it may be possible to discover their strengths and weaknesses which will, in turn, enable them to be better managed on the project. For instance, if a client tells you the subcontractor submits lots of claims, you'll know this must be managed during the contract to ensure they aren't given a reason to claim. You may even discover which their best teams are, and you can then request they are allocated to your project.

Bid shopping

Bid shopping is when you tell a subcontractor with a high price what the lowest bidder's price was. The subcontractor with the higher price is then given an opportunity to drop their price to below this, and you award the contract to them. This practice is not ethical and should be avoided. The higher priced subcontractor can be requested to drop their price (in which case all subcontractors should be given the opportunity to review their price), but they shouldn't be given the other contractor's price and told to beat it.

Labour and equipment supply subcontract documentation

The contract documentation must be clear and the rates should specify:
- rates for different categories of workers
- rates for different items of equipment
- what's included in the rates
- who will cover the mobilisation costs and the time for medicals and inductions
- if the rates for the workers include for:
  - overtime
  - leave pay
  - bonuses
  - allowances
  - personal protective equipment
  - insurances
  - special project allowances
  - accommodation
  - daily transport
  - food
  - lunch and rest breaks

- small tools
    - public holidays
    - supervision
- if the rates for the equipment include:
    - overtime
    - fuel
    - maintenance
    - repairs
    - consumables
    - lunch breaks
- the minimum hours that must be paid for in a day, a week or a month
- what happens in the case of inclement weather
- how disciplinary procedures will be applied to any of the subcontractor's workers

It should be clear that time sheets must be signed daily, with one of the contractor's staff members allocated this task.

The person responsible for supervising the contractor must understand the terms and conditions of the contract.

Adjudicating subcontract tenders

Subcontract tender adjudication is not as simple as it may seem, it's not just about choosing the lowest price. It's important to read through the subcontractors' tenders and quotations carefully to ensure that they have priced all of the works, used the correct quantities, considered all the drawings and priced the correct specifications. Sometimes the subcontractor may price an alternative product which they have highlighted in their covering letter or may simply have changed the specification in the bill of quantities they priced.

Trade and business competency must form a vital part of the adjudication process with the subcontractor experienced and skilled enough to deliver the project.

The adjudication of tenders must be seen to be done in a fair manner. If subcontractors believe their tender was not adjudicated fairly they may not tender for future projects, which would lead to fewer tenders, leading to uncompetitive bids, which may result in higher subcontract prices.

Subcontract negotiation

Subcontractor negotiation is not just about trying to get an additional discount from the subcontractor (although it is always worth trying to get a reduced price), it's an opportunity to ensure the subcontractor fully understands the scope of the project and the project conditions, and has priced their tender accordingly. In general, I would suggest that after going through the subcontract tender adjudication phase, during which the subcontractor has answered any questions the contractor may have regarding their tender, you call the subcontractor to a formal meeting with a set agenda which is circulated to the subcontractor ahead of time, giving them time to prepare.

This meeting should discuss:
- any unresolved issues and tender qualifications

- any specific queries relating to costs and items in the tender
- the subcontract schedule and method of construction, so the contractor is comfortable that the subcontractor understands the project, and can achieve the schedule
- the project scope, conditions and specifications, ensuring the subcontractor understands and has priced them accordingly
- how the subcontractor will staff the project, the availability of these resources, and their skills and capabilities

It's pointless, appointing a subcontractor who has under-priced the project, resulting in them losing money, and possibly leading to poor performance. In extreme cases the subcontractor may end up losing so much money that they go bankrupt before the end of the project, which will result in additional expenses and delays in sourcing another subcontractor.

Subcontractor documentation

The subcontract document is a legally binding contract between the subcontractor and contractor with enforceable provisions on both parties. If this document is poorly worded, inconsistent or incomplete, it can lead to complications with the management of the subcontractor resulting in the contractor incurring additional expenses, delays, even quality and safety problems on the project, and in the worst case, protracted legal arguments.

The subcontract contract document must include or make provision for:
- the scope of works
- the price
- contract conditions
- specifications
- applicable drawings
- specific site conditions
- a list of deliverables
- the subcontract schedule clearly showing commissioning and any discontinuities
- the safety requirements
- applicable quality requirements, procedures, tests and documentation
- commissioning requirements
- spare parts
- warranties and guarantees required
- clarification of what the subcontractor must supply to carry out the works, as well as the contractor's obligations

Much of the contract documentation should have formed part of the tender documentation.

Preconstruction (or kick-off) meetings

Before the subcontractor starts work on the project, I would recommend that a preconstruction meeting with a set agenda is held. If the subcontractor will be doing work on the project site this meeting should be on site, and the staff who will be interfacing with the subcontractor should attend, as well as the subcontractor's staff.

The purpose of the meeting is to:
- enable the parties to meet
- establish the lines and format of communication and where, and to whom, correspondence should be addressed
- reinforce the conditions of contract
- discuss the safety requirements
- go through the quality procedures
- enable the subcontractor to understand the logistical and administrative processes to mobilise staff and equipment onto the site
- provide the subcontractor with a list of relevant contacts
- discuss the schedule and methods of construction
- understand the subcontractor's requirements for services, laydown areas, offices and accommodation
- inform the subcontractor of the process, requirements and due date for the submission of monthly valuations
- ensure all parties understand the scope of works
- ensure the subcontractor is aware of the format, time frame and the channels for submitting contract claims such as delays, variations and extensions of time

Communication with project staff

Project staff must be familiar with the work that the subcontractor will be carrying out, as well as the responsibilities of both the contractor and the subcontractor. The person delegated to manage the subcontractor must ensure that the parties' rights and obligations are performed in a timely manner so that the subcontractor isn't delayed, has access to the work area, meets the required quality and safety standards, and complies with the contract specifications.

Subcontractor safety

Subcontractors must be aware of the safety standards that are expected of them. This must be conveyed in the tender documents, contract documents and at all subcontractor meetings. The subcontractor should have adequate safety systems and procedures in place and these should be assessed during the tender process. Poor safety performance should not be tolerated from subcontractors and their senior management should be advised of any breaches.

The subcontractor must have all the appropriate and required documentation including a complete safety file with all the relevant safety plans, procedures, registers, risk assessments and daily job start forms. These must comply with the project's safety standards and the approved project safety plan. If necessary the contractor may have to assist in putting these systems in place, since many subcontractors can be relatively unsophisticated regarding the project's requirements. This assistance may also include the training of the subcontractor's staff and workers. All the subcontractor's personnel must attend the contractor's safety induction.

Subcontractors with a large component of the overall project workforce should have workers elected to the project safety committee and also a sufficient number of trained first-aiders.

The contractor is responsible for the subcontractor's compliance with the project safety rules and procedures, as well as their legal compliance, with any accidents and incidents involving the subcontractor's personnel or equipment reflecting on the contractor's statistics reported to their client and to their management. The fact that the incident was not due to any action of the contractor, but was rather the subcontractor's personnel, will not be an excuse. Therefore, regular safety audits should be conducted on the subcontractor's safety systems to ensure they comply with the required standards.

Failure by the subcontractor to meet the required safety standards may cause the contractor's personnel to be endangered, and may result in not only the subcontractor's work being shut down (which will jeopardise the project schedule), but the whole project while the subcontractor remedies their faults.

Approval of staff and equipment

There should be a clause in the subcontractor's contract enabling the contractor to request they remove staff or equipment from the project if they aren't suitable, don't have the required skills or knowledge, or don't conform to the project rules and requirements.

Before the project starts, the subcontractor should provide the qualifications and experience of all staff they will utilise for the project. The proposed staff should have the appropriate knowledge and experience to deliver the project to the required standards.

The subcontractor's plant and equipment must also be appropriate for the tasks they will undertake. It should meet the safety requirements and standards required on the project since any that is in a poor condition, or poorly maintained, can result in safety or environmental problems, or breakdown resulting in the subcontractor losing time and falling behind their schedule.

Approval of subcontractors employed by the subcontractor

Subcontractors sometimes employ subcontractors to perform parts of their work. They may be contracted to manufacture equipment, fabricate sections, erect structures, or they may be specialists employed to do work which the subcontractor doesn't have the expertise to do. Before the subcontractor appoints their subcontractors they should seek the contractor's permission to use them, which enables the contractor to do checks and confirm they are capable of delivering the required work meeting the safety and quality expectations. The contractor may in turn have to seek the client's approval.

Approval of the subcontractor's subcontractor shouldn't detract from the subcontractor's obligations to perform in accordance with their contract and to ensure that their subcontractor meets such obligations.

The subcontractor is responsible for managing their subcontractors, and the contractor should direct all correspondence and instructions through the designated representative of the subcontractor, not directly to the subcontractor's subcontractor.

Joint ventures

Subcontractors should only form a joint venture to carry out their work if the contractor has given their consent. If it goes ahead there should be obvious lines of communication with the subcontractor, and it should be clear who the responsible partner is so there are no split responsibilities or confusion.

Subcontractor guarantees, warranties, sureties and insurances

The tender and contract documents should specify the guarantees, sureties and insurances required for the contract. Ensure these are received before work begins, and check that the wording and format complies with the contract requirements, and that the specified expiry date falls beyond the expected completion of the contract.

The Project Manager must check the name on the surety matches the name of the company you have contracted with. Often subcontractors trade under different names, and if the name of the entity contracted with is different to the one on the surety, the contractor may have no right to claim against the surety because the party named is not the one in default.

Also check the wording on the surety and the guarantees, since some wording can make it difficult to claim should the subcontractor default on the contract. If in doubt get an expert to look at it and provide advice.

Sureties, guarantees and warranties should be kept in a safe place. Warranties usually have to be handed over to the client at the end of the contract and payment may be withheld until they are received. Sureties have to be returned to the subcontractor once they have completed all their obligations on the project.

A register of guarantees, sureties and insurances should be established so the Project Manager can track which documents have been received and their expiry dates. Most insurance policies are bought for a period of one year, and before they expire a replacement policy should be provided. If the duration of the subcontract works are extended the guarantees may have to be reissued with a revised expiry date.

It's particularly important that the subcontractor has worker's compensation insurance to cover injury of their workers on the project.

In addition they must have suitable third party liability insurance to protect the main contractor and the client from any claims or lawsuits connected to the subcontractor's workforce or work.

Subcontractors who are designing and supplying items may also need to provide design indemnity insurance.

It's important to understand when the guarantee or warranty starts. For most items the period starts from when they are purchased. However, the subcontractor may keep the item in storage for several months before they install it, and it could be a further few months before the project is completed and the client takes occupation. By which time you may find the guarantee from the original supplier has almost expired. The client though, expects the piece of equipment is guaranteed for a specified period of time after they've commenced using it. This results in the main contractor being liable for any repairs the item requires between the period when the original supplier's guarantee ends and when the main contractor's guarantee period with the client ends.

Also, be sure to read the conditions of the guarantee because failure to adhere to special storage and maintenance instructions may render the guarantee void.

Communication with subcontractors

Communication with the subcontractor regarding variations, instructions, additional information, schedule changes and approvals, quality, safety and progress concerns, and all contractual matters should be addressed in writing to the subcontractor's authorised representative, and should emanate only from the contractor's designated representative. Any verbal discussions, regarding the above matters, should be followed up in writing to ensure there are no misunderstandings, there is a record of what was said, and that the appropriate people are aware of what was discussed.

Copies of all instructions and contractual information should be distributed to the Project Manager, as well as the contractor's contract administration staff.

Issuing drawings and instructions to subcontractors

Drawings and instructions must be formally issued to the subcontractor's authorised representative, and in reply a drawing issue slip or transmittal must be signed by the subcontractor and returned acknowledging receipt of same. These signed receipts must be kept on file by the contractor.

(Refer also to Chapter 4)

Instructions

Subcontractors must be dealt with in a professional and formal way, with any deviation from the original scope issued in the form of a written site instruction. Both the contractor's and subcontractor's staff must understand who is authorised to issue an instruction, as well as who is authorised to receive one. Failure to issue instructions in writing can lead to misinterpretations and confusion, possibly resulting in work not being carried out or being done incorrectly.

Subcontractor variations

Many subcontractors don't understand the procedure for submitting variations, often only submitting them when they have completed all their work. This is obviously unacceptable since many of them could be for the client's account, and they normally have a specified time within which they need to be notified. It's important that the subcontract documents include a variation clause which is back-to-back with the variation clauses in the main contract, but with a reduced time frame for the notifications and submissions, thereby allowing sufficient time for the contractor to scrutinise the claim and modify it with their mark-ups and rates, before it's submitted to their client.

The subcontractor must be aware of what details should be included in their variation submission, and to whom the variation should be submitted. The variation must detail:

- the reason for the variation

- the location of the variation
- when the variation occurred
- how the variation will be dealt with
- the cost and time impact of the variation

Subcontractor cost variations must relate to their contract document and the existing approved unit rates. If there are no approved unit rates, then the subcontractor must justify any new unit rates using the approved contract rates as a basis. Alternatively, the subcontractor must provide a breakdown of how the new rates were formulated.

If the subcontractor plans to do the additional works using day-works rates this method must be approved by the contractor before the work starts, with these based on agreed rates. The day-works sheets must be signed off daily by the authorised contractor's representative.

Subcontractors must also be aware of the approval process for claims and variations. Sometimes this is that the contractor must submit the claim to the client for approval, then, only once the client has approved the variation can the contractor approve the subcontractor's variation.

This process should be clarified in the contract document, and generally subcontractors should not begin variation work, or additional scope, without first obtaining written approval from the contractor.

Invoice procedures

Subcontractors must understand the procedures for submitting their monthly valuations, since they may have to be submitted in accordance with a specific format, including specific supporting documentation. Often these are submitted incomplete, to the incorrect person, wrong address, become lost, arrive too late for processing, or too late to be included in the contractor's claim to the client. This then causes aggravation to the subcontractors who won't be paid on time. Remember, a paid subcontractor is normally a happy subcontractor.

Subcontractor's shop drawings

The project and subcontractor's schedules must allow the time required to produce the drawings, as well as for the contractor and the client's representative to review and approve them.

(Refer also to Chapter 4)

Subcontractor progress meetings

Regular formal meetings should be held with all subcontractors, other than those doing only minor works. Minutes should be taken of these meetings and circulated to all parties as soon as possible. A standard template can be used.

Typically these meetings should include some or all of the following sections:
- safety
- schedule and progress
- manning
- mobilisation
- equipment

- quality control and test results
- outstanding information
- delays
- access requirements
- claims and variations
- monthly progress claims and payments

If, at these meetings, the subcontractor says there are no delays, they have access and there is no outstanding information, this should be minuted as such. I've often had subcontractors run behind schedule, and when I questioned them about this they replied they were delayed because they didn't have information or access to their work area. But having minuted there were no problems at every meeting we were able to refute these claims.

It should be mandatory that the subcontractor's authorised person, who is capable of making decisions, attends the meeting. Often subcontractors delegate their Supervisor or Leading-hand to attend the meetings, yet they are not authorised to make decisions, nor do they have all the information available to keep the contractor informed regarding progress, materials and resourcing.

Combined subcontractor meetings – interface meetings

Sometimes it's necessary to have interface meetings which are attended by various subcontractors. This is essential if one subcontractor's tasks are dependent on the progress of another's. With building construction there's always an interface between the different services which can result in clashes with the work-flow, as well as physical services. Some services are easier to install before others, while some services cannot be deviated and have to be fixed in a particular position and specified level. The purpose of these meetings is to resolve these clashes, and for the subcontractors to understand how their work impacts on others, and how others will impact on them.

These meetings should be minuted, and the minutes distributed as soon as possible. These should include who is responsible for actioning the item and the anticipated close out date of the action, this is important since other subcontractors will be planning their work-flow and teams around the expectation that the other contractors will have completed their required work ahead of them.

Payment of subcontractors

Subcontractors should be paid in accordance with the terms and conditions of their contract, within the time specified and correctly for the works they've completed.

Case study:

Subcontractors shouldn't be overpaid for the work they have done, yet this happened on one of my projects where the earthworks subcontractor had to excavate and backfill around various structures. Their rate covered for excavating and moving the material to a temporary storage location, and then, after the structures were complete, bringing the material back to backfill around them. However, after completing the excavation they claimed, and were paid, the full value of the item even though they still had to do the backfilling. When the time came to backfill the structures the subcontractor refused to carry out the work and defaulted on the contract. Although we still owed them some money, this was insufficient to cover the costs of employing another subcontractor to complete the works, and consequently we lost money.

Many contractors are keen to complete the easy work in their contract, and claim as much of their contract value as they can. Smaller works of low value often take far more effort and time, and cost the subcontractor more than they earn from completing them, so they may be reluctant to carry out these works, and may even default on the contract.

Often subcontractor payments are prepared and made by the Contract Administrators, or Quantity Surveyors, who sometimes work in isolation to the project staff. This is obviously unacceptable because they may not be aware of any problems that the project has with the subcontractor which may require set-off or back-charges. These problems may relate to poor quality work which has to be replaced, poor work performance which caused the main contractor to undertake some of the subcontractor's work, or the contractor supplying materials, equipment of personnel to assist the subcontractor to complete their work.

Retention monies must be retained in accordance with the contract document and payments should only be released once the subcontractor has handed over all the required sureties, bonds and proof of insurances.

Final payments should only be released once the subcontractor has fulfilled all the requirements of the contract. These may include:
- submitting the approved quality documentation
- handing over all guarantees and warranties
- completing all punch lists (both the contractor's and the client's)
- completing all the required commissioning and testing
- handing over the required spare parts
- handing over all operations and spares manuals
- resolving and agreeing all variations

I also strongly recommend that there's a formal final account document, or 'deed of final release', stating there are no outstanding claims and variations, that the subcontractor signs.

Back-charges

Contractors can be quick to back-charge subcontractors for work they have performed on the subcontractor's behalf. Yet the subcontractor may have had no prior notification that the contractor will be carrying out these works on their behalf, or that they will be back-charged the costs of them. It's important that the contract is followed in this regard, with the subcontractor being notified in writing and given the opportunity to do the works themselves.

Any back-charges must:
- clearly give the details of the work
- detail the documentation, reasons and notifications leading up to why the contractor undertook the works
- have the date the work was carried out

- substantiate the costs

Failure to properly follow these processes could lead to the subcontractor challenging the costs, possibly resulting in the contractor having to withdraw them despite incurring them.

Back-charges for utilities provided to the subcontractor should be agreed and signed off on a monthly basis, while equipment should be signed for daily. These rates should be agreed in the contract document, however, should there be no suitable rates in the document they should be agreed in writing with the subcontractor before the service is provided.

Clean-up work areas

Cleaning of work areas must be enforced regularly. Subcontractors should be made to keep their work areas clean and tidy, and dispose of all their waste material and packaging in the designated areas, or off site, as specified in the contract documentation.

Quality control

Procedures to monitor and control the subcontractor's quality must be in place from the start. These should be detailed in the subcontract tender and contract documents, so as to ensure the subcontractor is prepared and understands the quality requirements. The subcontractor must comply with the project-approved quality plan and documentation and the contractor should regularly audit the subcontractor's quality documentation to ensure compliance.

The contractor is responsible for checking the subcontractor produces quality work that meets the standards and specifications. Poor quality work from the subcontractor:
- will reflect on the overall quality of the project
- could result in the project being delayed while defective materials and workmanship is replaced
- may result in the contractor incurring additional costs if the subcontractor fails to rectify the work
- may result in the failure of an item or structure supplied by a subcontractor, and the client will hold the main contractor responsible
- costs them money, and if the subcontractor continually loses money due to their poor workmanship they could ultimately go out of business, which would result in the main contractor having to appoint another subcontractor at additional expense

If work is found to be substandard the subcontractor must be issued official notifications in the form of non-conformance reports, which they must close out as the work proceeds to ensure the non-conformance is rectified appropriately and that the problem does not reoccur.

Should the subcontractor be manufacturing equipment which requires factory inspections, the authorised and appropriate representatives from the contractor and client should be notified of these in advance to ensure the inspections are witnessed and signed off.

Samples, mock-ups and prototypes

It's good practice to request a subcontractor produce a sample, or build a mock-up, of the product they are supplying. In fact, this is often a client requirement which should be included in the tender and contract documents.

The purpose of a sample or mock-up is to ensure that:
- you and the client will be happy with the quality of the product
- the subcontractor is capable of producing the required quality
- there is a bench-mark for the quality standard for the project
- any interface problems or flaws with the design will be exposed
- it gives the client and professional team an opportunity to have a say in the finished product

Just a word of caution though, a sample or mock-up should not cause the client, Engineer or Architect to make fundamental changes to the specified product or finish, that will incur additional costs to either the contractor or the subcontractor. If this does occur the client should be informed there will be a variation. Obviously if there are quality problems with the original sample or mock-up, or the samples don't meet the required specifications, these are the contractor's and subcontractor's issues, and should be rectified at no additional cost to the client.

It's important, if the client or their representative has requested the sample or mock-up, that it's formally approved, either in writing or within the minutes of a project meeting. I've often had the client claim part way through the project that they never actually approved the item.

If possible, the sample or mock-up should remain to the end of the project, or at least until after the actual installed product has been accepted by the client. This should prevent arguments with the client over the standard of the final product since it should be the same as the approved sample.

Ensure the subcontractor makes money

Many may ask how it affects the project if a subcontractor makes or loses money. It would after all seem to be an anomaly, because if a subcontractor is making money it's probably at the project's expense? But no!

You must understand that a subcontractor who makes money is usually a happy subcontractor, and a happy subcontractor is:
- more likely to treat your project with the importance it deserves
- more willing to assist and accommodate the contractor when required

But a subcontractor who loses money:
- may cut staff and equipment on the project to save costs, which may jeopardise the project's safety, quality and schedule
- may take short cuts which could endanger safety and affect quality
- may use inferior, cheaper products
- often go to any lengths to lodge claims for additional costs and additional time, which if justified will cost the project money, however many will be spurious and merely an attempt to claw back some of the subcontractor's losses, yet they will still tie up the contractor's staff and cost money to evaluate and refute them

So why do subcontractors lose money and how can you assist them?
- If the subcontractor is producing poor quality work which has to be redone they will incur additional costs not allowed for.
- Many subcontractors may not have suitable staff or supervision on the project causing them to manage their works inefficiently, resulting in poor productivity and planning. Therefore assist subcontractors where possible to plan and manage their contracts as efficiently and productively as possible.
- Subcontractors may lose money because they don't have access to their work area due to the main contractor or another subcontractor delaying them. Many subcontractors will, of course, claim for these delays.
- Another reason for the subcontractor not making money may be because they are not maximising their revenue, although, of course, you don't want them doing this at your cost. Often though, subcontractors end up performing additional work for the client due to an increase of scope, or a change in the specifications, resulting in additional or more expensive items being required. If you're aware of any reason for the subcontractor to request additional monies from the client you should, in general, be encouraging the subcontractor to claim for these, and if necessary even assist them. If the claims are successful the subcontractor will not only earn more revenue, resulting in them making more profit, making them happier, but the contractor will usually earn additional profit as a result of their mark-up on the subcontractor's claim.

It should be remembered that many subcontractors may be unsophisticated in their approach to claims and may require assistance to not only maximise the claim against the client, but also to present it in such a way that it's not refuted by the client. By assisting the subcontractor with this you should also have some control in ensuring the claim is presented in such a way that your company, and project, doesn't risk being in the position where it has to pay monies that should be covered by the client.

However some subcontractors will never make money on a project, due to their incompetence, or a contract price that is too low. I certainly don't advocate you pay them money which isn't due to them. Still, it's important the Project Manager is aware of problems a subcontractor may have, particularly if they are losing money on the project, and takes certain mitigating actions, for example:
- tasks may have to be removed from the subcontractor's scope to ensure that work can proceed according to schedule
- assist the subcontractor where possible, and when unnecessary costs will not be incurred
- start preparing a replacement contractor (this may take the form of getting quotes for the outstanding work and checking on the availability of other subcontractors), to ensure minimal time is lost should it become necessary to appoint another subcontractor
- try to ensure that the subcontractor completes all work they have started so, should they leave the project, there aren't numerous unfinished tasks which will be difficult and costly to complete

- ensure major items of equipment and material already procured by the subcontractor are installed as soon as possible
- expedite the delivery of long-lead items supplied by the subcontractor
- ensure the subcontractor has not been overpaid, the retention money held is correct, and the bonds and sureties are valid

Signs of trouble

Often subcontractors may produce a product that is substandard, be losing money, be behind schedule, or have safety or industrial relations problems. If the Project Manager is working on a daily basis with the subcontractor they will often see the problem, days, weeks or even months before the manager of the subcontractor becomes aware of it. For this reason I would recommend the Project Manager takes a keen interest in the performance of the subcontractors, and raises any concerns they may have with their performance as soon as possible, so the subcontractor can implement steps to rectify the situation.

Termination of the subcontract and the right to vary the subcontract works

The termination clauses and the right to terminate or vary the subcontract works should be clearly stated in the subcontract document.

Termination should be used as a last resort for poor performance of a subcontractor since it will almost certainly cost the subcontractor money and their reputation, and they will usually not accept the termination without a legal fight. It also invariably results in the contractor incurring the additional costs of employing a replacement subcontractor. However, in the case of continued poor performance it will be a necessity, although it's often affected too late in the contract to really make a difference to the successful completion of the subcontract works.

The termination of a subcontract should not be taken lightly with other options being considered first; like negotiating with the subcontractor and possibly subletting portions of the work. The contractor's senior management should be advised before termination is instituted.

In many cases the termination is not done correctly which can then result in messy delays, and possibly even its withdrawal. If you've already appointed another subcontractor to carry out the works, and are then forced to reinstate the terminated subcontractor, there is the additional problem of terminating the replacement subcontractor.

It's therefore vital the subcontract document is clear on the right to terminate, and the reasons and steps that should be followed. Equally important is that the Project Manager terminates the subcontractor for the correct reason, checking all the facts, and ensuring the subcontractor has breached the subcontract agreement in a manner that allows for the contract to be terminated, and then carefully follows the steps in the document which outline the process to be followed to implement the termination. If the correct steps are not followed (including, but not limited to, the formal notification – using the correct form, quoting the correct clauses, addressing the termination to the correct person and to the correct address, ensuring the termination is formally received, and allowing for the correct notification periods), the subcontractor may be able to take legal action and recover all costs and lost profit.

Sometimes the contractor has to vary the quantity or type of work the subcontractor has been contracted to do. This may be due to the client varying the project scope, the subcontractor not performing resulting in the need for the contractor to reduce their scope, or because the contractor has revised their methodology. The contractor normally cannot remove scope from a subcontractor to give it unchanged to a cheaper subcontractor to do. The subcontractor will have to be formally notified if their scope is changed, with the reasons for the change. They may have to be recompensed for any costs they have already incurred on this scope.

Bribery and nepotism

Bribery, nepotism and favouritism must be condemned at all times. The contractor's and subcontractor's staff should clearly understand this, and appropriate action must be taken should any of these practices be discovered.

Many subcontractors may provide gifts to you or your staff. Some could be as simple as holding a function on site, tickets to football matches or other sporting events which shouldn't be a problem. Sometimes, however, they can be quite extravagant and could involve sporting events in other cities which not only include the entrance tickets but also airplane tickets and hotel accommodation. In extreme cases the gift can be even more substantial.

Expensive gifts can create problems, and even if it's not intended as a bribe it could be interpreted as such, and possibly will influence the way the subcontractor is dealt with. If you accept an extravagant gift your manager may question your dealings with the subcontractor and wonder if the subcontractor has been favoured as a result of the gift. A protocol should be established for gifts, and be included in the subcontractor's documentation with all staff aware of it.

Sometimes we become friendly with some of the owners or managers of subcontract businesses, or they may be related to you or one of your staff. All staff should declare any relationship they may have with a subcontractor working on the project so the Project Manager is aware of it and can ensure there are no conflicting interests. The Project Manager should also declare to their manager any relationships they have with a subcontractor. If a relationship may influence the subcontract tender adjudication process, the related party may have to excuse themselves from the final decision making process.

In extreme cases, you may find that a member of the contractor's staff has a direct interest in a subcontractor employed on the project. I've found in the past that some Supervisors and Engineers I employed owned equipment which they then hired to the project. This practice should be discouraged, because not only will that person ensure their business is treated in a favourable way, but it often happens that the individual attends to problems related to their hire business during work hours when they are being paid by the project.

Indigenous and local subcontractors

Sometimes projects have a requirement for a certain percentage of the work to be performed by local contractors and suppliers, or there may be a requirement that a certain percentage of the contract should be done by indigenous companies. There may even be a requirement in the country where the project is located for a portion of the work to be done by nationals of that country, or the indigenous community. Even if there's no legislation or rule apportioning some of the project works to these companies its good practice to consider using local or indigenous suppliers and contractors.

Often with new large projects, in remote communities, there is an expectation from the local companies that they will be given some of the work, although, sometimes their expectations of the amount of work and the profits may be unrealistic.

At the start of the project assess the capabilities of these companies and then consider what portion of the project can suitably be subcontracted to them. If the companies are very small and have limited capabilities consider requesting some of them to combine, forming a joint venture to undertake the work, or break the project into small packages suitable to subcontract to these companies.

It's important to remember that working with these companies may be hard work and time consuming. However, depending on the contract with the client, or the laws of the country where the project is located, there may be severe penalties for not subcontracting out the required percentage of work.

The Project Manager should ensure that their staff understands why work has been subcontracted to these companies. Since they often have to work harder to support them, if staff are unaware of the importance they will probably not help, which could lead to the subcontractor's failure. The contractor cannot afford for any of these companies to fail, since not only will they have to find a replacement but they will also face the prospect of being unable to reach the required quota of local subcontractors.

I've often, at additional cost to our projects, gone to great lengths to support smaller local and indigenous contractors with things like supplying supervision and training at little or no cost, and using our purchasing power to organise materials and equipment for them. We also paid them more frequently, and as soon as possible after they submitted their invoices, instead of paying claims monthly and thirty days after invoice.

The main purpose should be to train, grow and develop these companies so that when the project is completed they are capable of doing larger contracts. In fact, if you're really successful with one of these companies it may be possible, on the next project, to form a joint venture with them, making your bid more attractive to a client who is under pressure to give work to local or indigenous companies.

The subcontractor's expectations should be managed both in terms of the size of contract that is awarded and the possible profits. If they are expecting massive profits at the end of their contract they could be disappointed.

If a contractor is successful with assisting some of the local or indigenous companies it could pay dividends both in their dealings with the local community and with the client, and there may even be some political benefit. The Project Manager should remember that the contractor is often only working in a community for a short duration, however, the client is normally there for many years. Therefore if the contractor deals fairly with the local community it will be immensely beneficial to the client in the long run, and usually the client appreciates this effort.

Protection of work and the work of others

Subcontractors should ensure that their work is protected from damage by their workers, as well as by other workers on the project, and must also take care when working around other contractor's work. This should be included in their contract documents.

Payment of subcontractor's employees

It's the responsibility of the contractor to ensure that the subcontractor pays their workers on time and that the rates of pay comply with the regulations governing the project. Subcontractor's failure to pay their workers on time can lead to worker unrest which can spill over and affect other contractors on the project.

More than one contractor has been embarrassed either in the press, or by the client, when it has been found that one of their subcontractors employed workers at rates, or conditions, inferior to the regulated rates and conditions. It should, therefore, be included in the subcontract document that the contractor has the right to audit the pay and conditions of the subcontractor's workers. This process should happen at least once while the subcontractor is employed on the project, with the subcontractor immediately notified of any breaches discovered during the audit, or any complaints of conditions which are non-compliant. In addition subcontractors could be requested to submit a statutory declaration with their progress claims confirming they have paid workers, subcontractors and suppliers correctly on the project.

Maintenance and commissioning

The subcontract tender and contract documents must be clear on what maintenance and commissioning is required for the subcontract works, as well as what spare parts should be included. These should generally be the same as those specified in the contractor's contract with the client.

The schedule the subcontractor is working to should clearly indicate when the commissioning of their works will take place. Sometimes the subcontractor may be unable to commission their works immediately after completing their installation because the main contractor has other components to complete first. This may be because, for example, before the air conditioning system can be finally tested and commissioned the permanent power needs to be finished and tested, and the building needs to be complete. Knowing this the subcontractor should have allowed for returning to the project to carry out the commissioning in their contract price.

Ensure that the contract document is clear as to what types of testing and commissioning is required, who should witness it and how much notice is required. There may be factory testing of individual components, dry testing of individual systems on the project, and full wet commissioning when the whole system is completed and live.

The client normally requires commissioning documentation. This should be detailed in the subcontract document and be prepared ahead of the commissioning and while commissioning is progressing, to ensure it will not delay the handover of the section once it's complete.

The client's representative should be part of the commissioning process.

Signing off subcontract work

Subcontractors must be aware of the punch list procedure to close the contract out. These should be included in the tender and contract documents, the prestart subcontract meeting, and reiterated during the course of the subcontract works.

When the subcontractor has completed their work the contractor must inspect the works and draw up a punch list which the subcontractor must complete. But, where final commissioning is required, the subcontract work may only be able to be signed off after commissioning is complete. In addition, the work may have to be inspected and punch listed by the client, or their representative. However the norm here is that they usually only complete the final inspection of the subcontract works after the main contractor has completed the full scope of the project, which can actually be several months after the subcontractor has completed their installation.

When inspecting the subcontract works use the same critical eye as can be expected of the client, and ensure the subcontractor is aware of the inspection process, since invariably they expect when they have completed their project scope, and the contractor has signed off on these works, they have fulfilled their obligations. So when the client comes up with additional punch list items after they have completed the contractor's punch list, they are usually reluctant to attend to these, and in many cases may even expect additional payment to complete them. It's therefore good practice to ensure the documents allow for final payment to be withheld until the subcontractor has completed all items on the client's final inspection punch list.

Sign off work before the subcontractor demobilises to ensure it meets the quality standards and specifications. Often subcontractors are keen to return home, or they are required on another project, and in the rush to leave the project minor works may be left unfinished or punch lists left incomplete. It can then be difficult to motivate the subcontractor to return and close out their work as the majority of work is complete, and the major portion of their payment has been received.

Summary

- When asking for quotes and tenders from subcontractors ensure the subcontractor is provided with all the relevant documentation, including the terms and conditions of the contract, special conditions, drawings, scope of works, specifications and schedule.
- All the above should be included in the subcontract document when the subcontractor is appointed.
- Carefully adjudicate subcontractor tenders to ensure the tender complies with the tender document.
- Check that the subcontractor is capable of carrying out the work and has the resources and the capacity to deliver the project to the required quality and safety standards in the required time-frame.
- The subcontractor must be managed properly to ensure that the work is being carried out to the required standards and specifications, while keeping to the schedule.
- Subcontractor meetings must be held regularly with formal minutes kept of them.
- Delegate a responsible staff member to communicate with the

subcontractor.
- All instructions should be in writing and drawings must be formally issued.
- The subcontractor must submit the monthly claims in accordance with the contract and these must be checked before payment is made.
- The subcontractor must submit variations timeously and these should be approved and authorised before work begins.

Chapter 11 - Financial

Many Project Managers divorce themselves completely from financial matters leaving this up to Quantity Surveyors and Contract Administrators. I find this very strange because when I was a Project Manager I usually enjoyed preparing variations, measuring work, developing new rates, finding items not previously claimed, and doing the end of month claim or the monthly cost report.

Payment

Ensuring payment is received for work done is one of the most important financial responsibilities for the Project Manager on any project. After all, who works for free? However, many Project Managers leave payment issues to their Quantity Surveyors or Contract Administrators, who may not even be based full-time on the project, and consequently are often not always aware of the work done.

Many companies make good profits by managing their cash flows. They ensure the money owed to them is paid into their bank accounts on time, or even early, and they then earn interest on this money before having to pay it out to their suppliers and subcontractors. Unfortunately equally as many companies manage their cash flow poorly, resulting in them borrowing money from their bank to pay their subcontractors and suppliers, thus incurring additional interest costs. This poor management often means the company cannot finance the purchase of new equipment, and they may be unable to pay suppliers and contractors on time, resulting in unhappy suppliers, which makes the job of running the project more difficult. In the worst case the contractor may even become bankrupt and have to close down. Therefore it's imperative that the Project Manager ensures that money due to the company is paid on time, or ahead of time if possible, and that the revenue is maximised.

The person preparing the payment valuation should understand the contractual process and specified requirements, including:
- the date when the valuation is due, since some clients have strict cut off dates, and missing them could result in not being paid until the following month (I'm sure you would be very unhappy if you didn't receive your salary at the end of the month and had to wait another month!)
- the basis of the payment calculation:
    - some projects are only paid when a milestone is achieved and the Project Manager must understand what the requirements are for this, what must be completed and the paperwork required – I've seen milestones missed when Project Managers have not understood these requirements (sometimes 100% completion is not required, with only certain items having to be complete for the client to pay)
    - some projects pay in accordance with the percentage completion shown on the updated schedule, so it's important the schedule is updated correctly, as close to the required claim date as possible, with the person preparing the update aware that it will be used to

    adjudicate the monthly progress valuation
    o the project may be paid against measurement of the completed works, therefore this should be prepared accurately
- that in some cases the progress valuation is submitted ahead of the date of assessment, and it may be necessary to estimate the work that will be completed by the date of assessment
- the correct format to be used to present the valuation
- the supporting documentation that should be included with the valuation
- the variations and additional work completed in the month so it is included in the valuation

The Project Manager should review and check the claim before it's submitted.

  It's one thing to submit the claim, but equally important is to ensure the valuation is paid on time. Many clients and managing contractors regularly pay their contractor's monthly valuations late. This is sometimes because of administrative problems, but often I think it may be deliberate. After all if a contractor delays a ten million dollar claim for five days, this could be worth over five thousand dollars of interest (assuming a rate of 4%). Do this for twelve valuations and it becomes sixty thousand dollars!

  It's good practice to diarise when the payment is due, which may be a fixed date every month, or it could be a set number of days after the valuation or tax invoice was submitted. (In many instances, the client's representative must first certify the valuation and then only can the contractor issue their tax invoice with the certified amount. It's important that the contractor submits this invoice as soon as possible so the client doesn't have reason to delay the payment.) The day after the payment was due the Project Manager should check if it was paid, and if not the client's representative should be contacted to enquire where the payment is.

  Some clients are serial offenders and consistently pay late. There is always an excuse like, 'the invoice wasn't received on time,' 'the person who authorises the payment was on leave,' 'the paperwork was lost,' 'the contractor did not submit the correct paperwork,' and so on. In this case the Project Manager should be reminding the client several weeks before the payment is due and follow up where the payment is in the client's system so there are no excuses. In addition, ensure the issue of late payments is minuted in the client's progress meeting and recorded in emails. Sometimes a call to a senior member of the client's team may unblock the payments especially if payments are deliberately delayed by a junior administrator who doesn't understand the importance of paying a contractor on time.

  If nothing works, and the contractor is continually paid late, a formal letter should be submitted to the client advising them that late payments are unacceptable, not in accordance with the contract, and that interest will be charged (this assumes the contract allows the contractor to claim interest for late payments). Most clients don't like paying interest on overdue payments.

Payment for unfixed materials

  Many clients don't pay for materials which have been delivered to site but haven't been built in. Yet, the contractor still has to pay the supplier for the material. To overcome this, the contractor should minimise the quantity of unfixed material by installing it as soon as possible, preferably before the due date for the monthly valuation.

It's worth noting that some clients will pay for the unfixed material, but, only providing certain conditions and paperwork are completed and submitted with the claim. The Project Manager must ensure this is done correctly.

Payment retention

Many projects require cash retention to be withheld by the client, with it only being released once the contractor has successfully met various conditions and milestones. The Project Manager should be aware of these milestones and their requirements, and ensure they are met as soon as possible, and that once they are, steps should be taken to obtain the release of the retention.

Often the final retention money is only released at the end of the maintenance period. In many cases this may be a year or more after the completion of the project, after the client has done a final inspection of the project, and the contractor has attended to all the defects that were discovered during the inspection. It's important the Project Manager arranges these inspections well in advance of the due date, and then attends to the defects speedily, enabling the earliest release of the retention money.

Non-payment

Non-payment probably results in more contractors going insolvent than anything else, so to avoid this it's vital to follow-up and ensure the company is paid on time. Equally important is taking cognisance and investigating any rumour or references to the client being unable to pay other contractors, or any talk in the market place regarding the solvency of the client. If there are any suspicions surrounding the client's possible cash flow problems:
- report them to senior management
- investigated them further
- where there's merit, or basis, to these rumours consider calling a meeting with the client to discuss them
- refer to the contract and check what recourse there is should the client be unable to pay
- delay the delivery of non-essential items
- avoid placing new orders for material and equipment where possible
- remove equipment from the site which isn't being used
- where possible, slow down the rate of work without compromising the contractor's rights and obligations
- if possible avoid mobilising or employing new personnel
- check if a payment guarantee is in place, and that it's valid and will cover the work that has been completed but not yet paid

If the monthly valuation has not been paid, and there are reasons to suspect that the client may have cash flow problems, it's important the Project Manager reacts quickly, but in such a way that doesn't jeopardise the contractor's contractual rights. It should be noted that if the client has insufficient money, every day the contractor continues working on the project is another day's work that may not be paid. It's an amount that can quickly escalate.

The Project Manager must understand the contract since in most cases the contractor can't just stop work on a project because of non-payment. Stopping work could result in the contractor being in default on the contract, which could give the client the right to terminate the contract, giving them a legitimate reason not to pay the outstanding money.

Day-works or cost-plus-fee

Sometimes the company is fortunate and the contract is paid on a cost recovery, a cost-plus, or day-works basis, in other words payment is made for the hours worked (both for the workers and the plant and equipment) at previously agreed rates, and for materials supplied at their cost plus a previously agreed mark-up. What a pleasure! The contractor is paid for their actual costs and the hours worked, plus a profit margin on top of this. What could be better? There's no risk, and most contractors dream of these projects. You cannot lose!

Believe it or not, you can lose. Why does it happen? Well, normally because the Project Manager has not performed his job properly.

Case study:

Recently I became involved in a project that in the previous five months had completed over ten million dollars of work, done on a cost-recovery basis, but they had only invoiced for half a million dollars! Hard to believe! Imagine what that did to the project's cash flow. Even harder to believe was that none of the hours worked by people or machines were signed for by the managing contractor. Not because they didn't want to sign and agree them, but simply because there were no daily records kept. Furthermore materials purchased for the works were not agreed, and copies of all the expenses and invoices were not readily available.

It took us months to try and resurrect all the back-up paperwork, and to put the hours that people and machines worked onto daily records, which we then had to get signed off by the managing contractor. Fortunately, they agreed to sign the records, although many other contractors could have refused so long after the event. How much simpler it would have been if the records had been prepared on a daily basis, when everything was fresh in people's minds, and then signed and agreed the next day. Equally as simple, as the invoices for purchased items being printed and submitted with the day-works records. Done this way at the end of each month everything should have been agreed and invoiced.

Apart from the problem of the negative cash-flow, who knows what records were lost in the interim, and what work ended up being done and never invoiced for.

There are a few basic principles involved in doing work on a cost-plus, cost-recovery, or day-works basis.

- Ensure accurate records are kept of the work performed. This is not necessarily as straight forward as it sounds. Many Project Managers leave this up to support staff such as an Engineer, Supervisor or the Site Administrator, despite them not entirely understanding what they are doing, and the importance of producing accurate records. I worked on a project where the Supervisor was completing these records and the Project

Manager hadn't even explained to him that the records were used as a basis for a variation from the client. The Supervisor later told me, if he had known how important the records were he would have ensured they were far more accurate and complete.
- It's essential to understand what is being paid for and what is included in the rates for people and equipment. For instance, does the equipment rate include fuel, the cost of replacing wearing parts and tyres, lubricants, repairs and maintenance? Does the rate for the people include for tools and personal protective equipment? If any of these items are not included in the rates then separate records need to be kept for these costs.
- When claiming time for people, plant and equipment, ensure the full hours are claimed. On many projects a worker may, for example, work a ten hour day and get paid for ten hours, yet the worker actually only worked, say, seven hours on the task for which the contractor is being reimbursed. So what did the worker do for the rest of the day? Well, he probably attended a pre-start meeting, then collected the item of equipment necessary for the job, did a pre-start check on it, moved the item to the task area, prepared a job hazard assessment and completed other relevant paperwork. During the course of the day he would also have had approved rest breaks. In other words, in the course of the ten hour day the worker had only worked on the one task, and the full ten hours should be claimed and paid.
- Before performing work, on a cost-plus basis, certain principles must be agreed with the client or managing contractor:
    - Who from the client will be signing the daily records, and are they authorised to approve these records (it's pointless having records signed by a person who doesn't have the required authority)?
    - The agreed basis of payment:
        - the hours of work
        - overtime rates
        - payment for management and supervisory staff
        - what happens with staff that may be working on the project off site
        - what happens with staff shared between different tasks, such as Site Administrators, Safety Advisors and Contract Administrators (agree on the proportion of these people's time that will be paid)
        - the rates for equipment and plant
        - the cost of repairs and maintenance
        - the transport of the equipment to and from the project
        - the cost of the service and refuelling vehicles
    - Agree with the client on what types of records are required (for instance, are delivery notes sufficient for materials or are invoices required).
    - Will the client pay for consumable materials like glue, nails, bolts and cable ties?
- Understand what materials the client will supply for the works.
    - What happens if these materials are not supplied on time? Can

- alternative arrangements be made to purchase the missing materials and will these costs be reimbursed?
  o Items supplied by the client may have to be offloaded, stored on the project and then moved into position and these costs must be invoiced to the client.
- Often materials are taken out of the contractor's general store, with no record kept of these to enable their costs to be invoiced. Even if there is a record, it's often difficult to justify their cost, as there may not be separate invoices for them. Where ever possible material should be specifically ordered for the cost-plus-task and personnel must be informed which material is to be used for the task.
- A cost that is often not captured and invoiced for is the offloading of the materials, because this is usually done by a team not directly involved with the task.
- Include the costs for insurances, bonds, transport, travel, accommodation and royalties, if these are applicable.

Case study:

We had a project that the owner of the business priced and managed. He told us the contract was priced on a cost-recovery basis with a 15% mark-up. We were, however, only making 11% so I investigated the problem.

- *I found the agreed mark-up for materials was only 10%, not 15% as per the rest of the contract.*
- *We had a fuel and service vehicle operated by a fitter which the client refused to pay for as they maintained this was allowed for in our equipment rates, and we hadn't specifically excluded this from these rates.*
- *The client also refused to pay for equipment that was not being used. We should have specified in our contract that standing time rates would apply to items on site if they were not used. In fact, sometimes the reason the items were not being used was because the client had specified the number and types of equipment, as well as the number of operators – and the number of operators was always less than the items of equipment. Yet, we had to pay our suppliers for the equipment on the project regardless of whether the item worked.*
- *During the course of the project there were several days we were unable to work due to rain, and the client didn't pay us for this. However, we still had to pay our workers and the equipment suppliers for these days.*
- *The time our Project Manager spent on the project was viewed by the client as a Head Office overhead, so we weren't paid these costs.*

From the above it can be seen how important it is to ensure that any cost-recovery contract is carefully set-up, and that all costs are taken into account when preparing and agreeing the rates with the client.

Another possible problem to be aware of is that the client may issue a cost-plus contract, or variation, including an upper limit to the contract value. If this limit is exceeded there's a risk that the amount in excess will not be paid, because the client may not have budgeted for the additional money and consequently may not have the funds. The onus is on the contractor to keep the records of the costs up to date, ensuring they are complete, and to review the total costs of the works regularly, advising the client in advance if the value in the variation will be exceeded. Bear in mind that the client could take several days, or even weeks, to issue a revised contract or variation, and if there is a delay, work may be halted, resulting in additional costs being incurred, which may not be for the client's account.

Anyway, it's good practice to regularly update the client with an accurate estimate of the costs incurred. The client will be unhappy if a particular value is consistently reported, and then when the final invoices are collated the value has suddenly jumped to a much higher figure.

Despite what I have said above, I've had many profitable cost-recovery projects, and have in most cases made a profit which exceeded the mark-up percentage.

Variations

Variations are normally a result of a change on the project, and could be due to:
- changes in quantities
- revised specifications
- altered working conditions on the project
- additions or omissions of items and sections of work
- the client not providing what they should have, or providing it late
- delays

With lump-sum price contracts it's particularly important to compare the construction drawings with the tender drawings and the tender scope of works to see what has changed. In addition it's useful to compare the tender specifications with the construction ones. Any differences which increased the contractor's costs should be claimed as a variation.

Many Project Managers leave the formulation of variations to their Quantity Surveyors or Contract Administrators, however I believe the Project Manager should play a major role in the formulation of variations and no variation should be submitted without their knowledge.

(Refer to Chapter 12 for lodging claims)

Costing variations

The costs of carrying out variation work may include:
- the cost of materials, including:
    - transport
    - unloading
    - moving them into position
    - all fastenings and fixings
    - wastage
- the cost of supervision, including:
    - salaries
    - transport
    - accommodation
    - leave pay, benefits and bonuses

- overhead costs which may include:
    - salaries of the managing and administrative staff
    - the provision of offices
    - site facilities
    - safety equipment
    - security
    - additional bonds and insurance premiums
- labour costs, including:
    - the actual hourly rate
    - transport of the workers
    - accommodation
    - overtime premium
    - personal protective equipment
    - leave pay, paid public holidays, benefits and allowances
    - non-productive time (time for inductions, mobilisation and preparing the work area)
- any rework of completed work which may be required to enable the new work to proceed
- protection of completed work to ensure it's not damaged when the new work is done
- equipment costs, including:
    - the hourly rate for the item
    - cost of the operator
    - fuel
    - maintenance
    - any unproductive time
    - consumables, cutting and wearing edges
    - cost to transport the item to the site and then to remove it
- the cost of any formwork, scaffolding and access equipment
- the cost of being longer on site
- the impact on other tasks on the schedule

All costs in the variation must be justifiable and provable to the client. Some Project Managers like to submit variations with a higher value of what they are ultimately looking for, in the expectation the client will knock their price down to a value close to what they actually want. I personally don't like doing this because you lose credibility with the client who gets the impression you are deliberately overcharging them. I would rather present a variation cost that is accurate, which the client has difficulty in finding fault with, and will have to agree to pay.

Change orders

Normally once a client has accepted a variation they have to issue a contract amendment or change order, since the contractor will normally not be paid for variation work until one is issued. Therefore the Project Manager should track variations, and ensure the appropriate change orders are issued so payment is not delayed. I've often waited several months for change orders to be processed.

Change orders should be in writing and you should check:
- the value is correct

- the written conditions are acceptable, since I've had clients add in completion dates on the order, which were not previously agreed
- they are signed by both parties
- they clearly define the scope
- they specify the time in which the works must be carried out

Cost reporting

Generally, most Project Managers work for a contractor who has an established financial software package in place, and a standard format for cost reporting which must be submitted monthly, or in some instances, weekly.

Hopefully the system the company uses is not too involved or complicated since there is not much you'll be able to do to change it. I've worked for companies with monthly reports nearly a hundred pages long and alongside others with cost reports of thirty pages going into detail on every aspect of the costs. It's worth keeping a few things in mind:

- The time spent preparing these lengthy reports can often be put to better use elsewhere on the project.
- There is often so much data to be assembled for the report that the project may use the incorrect details, or data that has not been interrogated properly. Unfortunately the old adage of 'garbage in equals garbage out', definitely applies with cost reports – and many of them are rubbish.
- Lengthy reports, with too much information to be digested, are often not read, or if they are people don't focus on the important data, and items to be acted on.

Case study:

One of my projects was in an African country in joint venture with another large civil contractor who was the lead partner. The project used their financial reporting and costing system which consisted of a lengthy cost report.

The project required about twenty thousand cubic metres of concrete, which was produced from our concrete mixing plant on site. The plant used cement which we transported, using two of our own cement tankers, from a cement factory six hours drive from our site. The project was constantly delayed due to a shortage of cement.

There was a monthly joint venture meeting at which we went through the lengthy cost report. In the second month, I noticed we were losing money on concrete materials, and when I queried this, I was told the client had instructed us to use additional cement, the costs of which we would be claiming from the client. The next month the loss on concrete had increased, again I queried this, only to be told when we submitted the variation for the additional cement we would recover the loss. I asked if we were reconciling the cement on the project and was informed this hadn't happened, but the project management team would attend to it.

Concrete material reconciliation is fairly simple, and should be done regularly on all concrete projects. It's done by working out how much concrete has been placed into the individual structures and comparing this with the amount of concrete supplied from the project's concrete plant, or from a ready-mix supplier. These quantities should match, ensuring the ready-mix supplier has supplied the quantity of concrete they have invoiced, and that the project has not wasted concrete, by for example, placing the concrete thicker than it should be.

If the concrete produced from the project's plant is known, it's possible to calculate the quantity of materials (cement, stone and sand), that should have been used to produce the concrete, and to compare these with the quantities actually used. Any shortfall would either be due to the project wasting the materials, or the supplier charging for materials they had not actually delivered.

A couple more months went past and the loss on concrete materials got worse, with the same excuse used to explain it. I eventually got a chance to visit the site, and due to the infrequency of flights, I had to spend several days there, so had time to do the concrete material reconciliation myself. When we calculated how much cement we should have used, and compared it with the quantity we had actually paid for, we discovered there was more than half a million dollars of cement unaccounted for.

On investigation we found that when our drivers returned to the site with a load of cement they made a detour, stopping off and discharging the cement from one of the tanker's compartments, which they then sold. In fact, a third of every load of cement was being stolen from the truck en-route to the site.

Not only was there the direct cost of the stolen cement, the project also always had insufficient cement due to every truck only delivering two thirds of what they should have, and the trucks taking a couple of hours longer on each return journey because of the detour to unload cement.

By implementing suitable controls we were able to prevent further theft of cement, but obviously we were never going to recover the losses we had already incurred.

The lessons from the above are that the Project Manager should:
- implement controls and systems to minimise material losses because they can significantly impact the project
- reconcile major materials regularly
- act on, and interrogate the reason for a particular loss reported in the cost report

Cost reporting is normally done by allocating items on the project to cost codes. The cost report compares the cost of the item, or task, against the allowable, which is the income or revenue earned by the item or task. Contractors are normally interested in analysing labour, materials, equipment and subcontractors, while a more detailed analysis may break each of these items down into further subsections.

Cost reports should take into account the effects of upfront payments, over-claims and the payment of overheads. Failure to take correct cognisance of these factors could result in the cost report being incorrect.

The reports are only as good as the quality of the information entered into them. Therefore the Project Manager should take an active part in managing the process, to ensure the report is accurate, and then use the information produced to maximise the profits and minimise any losses.

Cost to completion

Cost to completion is a good method to accurately forecast if a project will make or lose money. It involves taking the costs incurred on the project to date, and adding to them all the costs that the contractor expects to incur to complete the project. The estimation of the additional costs is done by extrapolating the costs incurred to date, for the individual cost codes, by checking and adding what costs can be expected to be incurred to complete the remaining work.

Obviously the accuracy of this method depends on the Project Manager ensuring the costs to date are accurate and complete, and that all possible future costs (both direct and overhead costs), have been included in the estimate. The larger cost items should be done more accurately, as a miscalculation on just one can have a significant impact on the overall estimated profit or loss.

Project budgets

It's important an accurate budget is prepared for the project and that it's updated regularly during the course of the project. This budget is a forecast of the potential financial outcome of the project and an indication of possible financial problems which may be encountered. Project budgets are normally required by the contractor's management to help predict the company's overall financial position.

The Project Manager must ensure the budget is prepared properly, and all potential costs have been taken into account. If an accurate budget is prepared the final project financial position should be very similar to the estimate.

Any variations from the expected budget estimates in the cost report must be investigated and explained by the Project Manager. As important as it is to explain negative variances, I would also expect the positive variances to be verified and explained to ensure they are correct.

Cost codes

Most companies develop their own set of cost codes which are used on all their projects. I don't want to dwell too much on the subject but there are a few important items.

- Cost codes should be simple and easy to use. The more codes there are, and the more difficult they are to use, the less likely it is that personnel will use them correctly, which will result in items being allocated to incorrect codes, resulting in an inaccurate cost report.
- They must be allocated from the start of the project and should be used by everyone associated with the project.
- The correct use of cost codes must be strictly enforced and if necessary all personnel using and allocating cost codes should attend a brief induction explaining how they work, and the importance of using them correctly.

If the costs and revenue incurred by the different tasks and items can be correctly allocated to a cost code, it's possible to get an accurate portrait of whether that task, or item, is making or losing money, enabling the appropriate steps to be taken to rectify any losses.

Chapter 11 - Financial

Trading on claims

Trading on claims is when a contractor submits a variation to the client, or thinks they have reason to formulate a variation, and they then use the revenue they assume the client will pay them for the variation as revenue in their cost reports. This is a dangerous practice since it firstly assumes that the variation is legitimate and the client will agree to it, and secondly, that the client will agree to its value. In many cases this is not the case, which leads to the cost report being overly optimistic declaring a larger profit (or smaller loss), than the final project result.

Trading on claims gives the project team a false sense of comfort as they believe they are doing better than they actually are.

I would much rather report a lesser profit (or even a loss), and then when the value of the variation is finally agreed with the client increase the profit accordingly.

Contracts that are losing money

If the contract is losing money what do you do? Many Project Managers try and hide the fact by trading on claims, over-claiming their monthly valuations, or not declaring some of the costs, all of which are dangerous practices since you're trading with money you're not entitled to.

What do your managers want you to do? Most will say cut costs, cut people, cut equipment, cut anything to reduce costs. Certainly in some cases this may help especially when the project is over-staffed or equipment is used inefficiently.

However, if the project is showing a loss the most important thing is to establish the reasons. I say 'reasons' because there are often a number of causes for the loss. In many cases the project finds one reason – for instance, there's one major item under-priced in the tender, which they then believe is the only reason for the loss, and for the rest of the project you hear how a poorly priced tender is causing the project to lose money. Yet, it probably wasn't the only reason the project lost money, it was probably just one of the reasons.

Case study:

On one of my projects we were in joint venture with other contractors constructing a dam. It was valued at over a hundred million dollars and took several years to build. Over the course of the project the project team were consistently falling behind the schedule and were continually requesting more resources, more people, cranes, concrete trucks, concrete mixing equipment, in fact more of everything related to the concrete work. These additional resources were not allowed for in the tender, and were definitely not budgeted for, resulting in us reporting a loss.

The project was priced on quantities supplied by the client. These quantities were subject to re-measurement and the contractor was entitled to be paid for the final measured quantities. However, during the course of the contract we hadn't kept this measurement up-to date, and it was only towards the end of the project that we found the final quantities for the concrete work, in many instances, exceeded the tender quantities by more than 25%. Despite these increases we still managed to complete the project within the project schedule. What we had, in fact, done was accelerate, so the additional work was completed in the same time, which is why the project had required all the additional resources, but we hadn't claimed for this acceleration. Once we knew this, we were able to claim for these resources, and eventually, long after we completed the project, were paid for them, enabling the project to show a profit.

What should have been done, at a much earlier stage, was investigate the reason why the additional resources were required, so we could lodge the claim earlier, which would have helped our cash flow, and perhaps even have put ourselves in a stronger position to settle the claim for a higher value.

Of course, if we'd had managers who were steadfast and refused to resource the contract beyond that budgeted for at tender stage, then we wouldn't have completed the project on time.

Losses on a project may be because resources are used inefficiently which may be due to:
- the project having too many people
- there being insufficient supervision
- the wrong mix of trades, as in too many of one trade and too few of another trade (for instance if there are insufficient scaffolders they may not be able to build scaffold fast enough which could hold-up bricklayers, reinforcing-hands or form-workers using the scaffold)
- the project having too many items of equipment
- items of equipment that keep breaking down
- items that are incorrect for the task (for example, excavators that are too small resulting in longer loading times)
- the wrong mix of equipment (for instance, too many trucks for the number of excavators)
- the project being poorly managed and planned
- the workers being poorly trained
- the workers being poorly disciplined, resulting in a poor work ethic
- poor quality of workmanship, resulting in rework
- defective materials having been used
- subcontractors not performing as they should, delaying follow on works
- there being a poor attitude amongst the workers, or a genuine grievance or problem causing unhappiness and affecting the productivity

Of course, the reason for the loss may be because conditions are different to those anticipated at tender stage, resulting in poorer productivity from people or equipment. Sometimes it's because the Estimator made the wrong assumptions about the conditions, however, sometimes the conditions are different to those expected because something has changed. These differences could be any number of issues including:
- the client imposing additional operational conditions, for instance:
    - imposing lower speed restrictions than normal
    - limiting the work hours
    - changing specifications
    - changing testing regimes
    - changing haul road routes
    - locating services further from the work area than was provided for at tender
    - moving spoil and borrow sites further from the work area

- there could be different environmental factors:
    - the ground could be harder than anticipated
    - there may be rock
    - the borrow areas may not be accessible
    - the available material may have different characteristics to those specified

Case study:

One of my projects was a bulk earthworks project, and at the time of tender the client had specified where water would be provided, and we subsequently provided the client with our expected daily water demands. When we started the construction the client had not completed the water supply point specified in the tender, so we had to travel an additional four and half kilometre to the nearest supply point. Because of the additional distance and travel time we needed extra tankers to transport the water. In addition to this, the tankers had to queue because other contractors were also using the water point, and there was limited water available, so it took longer than expected to fill the tankers.

Now those of you with earthworks experience understand that water is required to compact the material placed in the roads and terraces being constructed. The consequence of having insufficient water was that we were only able to process about three thousand cubic metres of material every day, instead of the required five thousand cubic metres. This resulted in our personnel and equipment not operating at the efficiencies they were required to operate at to cover their costs. It also resulted in the project taking longer to complete.

We had to submit a claim for these delays and additional costs, but unfortunately in this case we were not entirely blameless, since we had our own inefficiencies, which didn't help our situation and made our claim difficult to justify.

The client could be causing the inefficient use of the contractor's resources by delaying the project due to:
- issuing information late or of a poor quality
- not meeting the access dates
- taking longer to approve designs and drawings
- issuing instructions changing the scope or resulting in rework

If the client has caused a delay, or there are changed conditions from those mentioned in the tender and contract documents, it's usually possible to submit a claim to the client for the additional costs and time.

Sometimes the loss could be simply because the contractor hasn't claimed for additional work they have done, or for other variations they are entitled to claim under the contract.

Weather can impact both productivity and progress on a project. Most often there is little that can be done to prevent this problem. However, the project manager should understand the conditions included in the contract document and, depending on these conditions and the type of weather event, there may be an opportunity to claim from the client or against insurance. Sometimes the impact from the weather may be as a result of a shift in the schedule, which may be due to delays created by the client, resulting in certain tasks being performed during a season they were originally planned to avoid. Occasionally, it's possible to do the works differently, to use different equipment, or try and work around the problem. These revised methods may entail additional costs, but may be less than those directly associated with the weather's impact.

Whatever is causing the loss, it's important the Project Manager finds the causes, and takes steps to rectify it or them, to prevent further wasted costs, or to find and claim additional revenue to cover the costs causing the loss.

Basic site costing

The project should be completing a monthly cost report as outlined above. This report, however, is often dealing with information a month old, so it's useful if the Project Manager and Section Engineers practice some basic daily site costing to help detect problems at an early stage. These calculations can be fairly simple and used to track a few repetitive tasks, which can add up to a significant portion of the project's cost. For instance, if you have an excavator loading trucks, it's very simple to work out the hourly cost to operate the excavator and trucks (the hire rates, plus the operators, fuel and maintenance), and using the allowable (what the client is paying you, less mark-up, to excavate and move a cubic metre of ground), you can quickly calculate how many cubic metres of ground must be moved in an hour to make money. If more ground is being moved than calculated, then the project should be making money, while if less ground is being moved there is a problem and either the costs must be reduced, (for example using a smaller excavator or removing a truck), or the production must be increased to achieve the required quantity.

A similar process can be performed for most operations on a construction site whether it's the tonnes of reinforcing fixed, cubic metres of concrete poured, or metres of cable installed in a day. The results don't have to be 100% accurate, but they should be accurate enough to provide a quick, and reasonable, indication of whether the task is making or losing money.

A word of warning though, these calculations should preferably cover a full shift's production. I have seen Project Managers check the production done in a hour and think the work is making money, forgetting however, that in the course of the day there are times of no production, for instance, when the team moves between tasks, during servicing and refuelling of equipment and the lost time on either side of the workers' breaks.

Buy in from staff and feedback

Information obtained from the cost reports should be used to give feedback to the project management staff and Supervisors. I don't mean distribute a copy of the cost report because this often contains sensitive information, however, they must be told where they are losing money, and should be congratulated for those tasks they are doing profitably. Many Supervisors think their teams are performing well, and are surprised when they are told they are losing money, and once aware of the problem they may implement steps to rectify the situation.

Reconciliation of materials

It's important to reconcile the materials bought on the project with the materials invoiced to the client – hopefully these amounts will agree. If the quantity

of materials invoiced to the client is less than that purchased it could be due to:
- wastage through cutting, bulking, breakages, or lapping of materials
- the incorrect quantity being invoiced to the client, which may be due to measurement errors or additional work and variations being overlooked
- theft which cause additional costs and delays, and may happen:
  - on the project site
  - in the delivery process
  - because the material may never have left the supplier in the first place

Case study:
One of the concrete silos we constructed had about two hundred cubic metres of concrete in the walls which was supplied by a ready-mix company. When we completed the silo my Project Manager reconciled the concrete supplied, with the concrete that was actually in the walls, and found we had been invoiced for two hundred and twenty-five cubic metres, which was 10% more! We called the manager of the company to our offices and asked for an explanation. They first claimed we had wasted the concrete, then that we had returned trucks with concrete unused, and then claimed we had formed the walls thicker than we should have. We were able to refute all these arguments and they eventually agreed to give us credit for the excess concrete. We will never know if this was a genuine error or if it was an attempt to be paid for concrete that was not supplied.

Of course twenty-five cubic metres of concrete is not that much, but it could just as well have been 10% of ten thousand cubic metres, which would have been a very large additional cost.

This 'missing' concrete, cement, stone, and sand happens so often that it's sometimes hard to imagine it's always a genuine error. Of course, errors do occur and occasionally I've had the opposite happen, when we received deliveries with a larger quantity than we were charged for.

By reconciling the material deliveries on a regular basis it should be possible to take action to prevent further losses, or to invoice the client for the items which were overlooked.

Payment for goods and services

The Project Manager should check that suppliers are paid correctly and on time. There must be a system in place to pay for goods and services, which matches up the quantities, rates, and types of goods delivered, with those ordered, and those invoiced. This will negate the errors with invoicing which can lead to suppliers being overpaid.

Goods and services should be paid for in accordance with the provisions of the order, since interest may be levied on late payments or payment discounts forfeited.

In addition, suppliers who are regularly paid late will not be willing to do business with the contractor, or may they may even add a factor onto their prices to allow for the late payment. Normally suppliers paid on time will be more obliging in supplying an efficient service and will accommodate changes to orders and assist with emergency orders.

Vehicle log books and expense claims

Vehicle log books and expense claims should be completed monthly, authorised by the Project Manager, and submitted to Head Office for payment and incorporation into the project costs. The claims should have all the supporting information attached and be cost coded correctly. Often I've had Project Managers who were lax with completing this paperwork, resulting in the claims being submitted months late, sometimes even after the project was complete. This put the individual's personal finances in stress, and also resulted in the late costs impacting adversely on the project cost report.

Expense and travel claims which are not checked properly can result in personnel claiming expenses that they were not entitled to, resulting in additional costs to the project.

Summary

The Project Manager should:
- ensure the monthly valuation claims are prepared correctly and submitted to the client in the correct format, with the correct paperwork, and on the due date specified in the contract
- ensure the client pays the valuation within the specified time frame
- monitor the project's cash flow since a negative cash flow can impact the contractor's ability to trade successfully
- ensure variations are submitted within the time specified in the contract document
- should review all variations, checking they have considered all the costs and the claim has been formulated correctly, referencing the correct reasons for the claim, quoting the necessary drawings and contract clauses, and including all the relevant calculations
- ensure the client issues a change order for variations
- prepare a project budget
- when undertaking work on a cost-recovery contract, ensure all the costs for this work are included, gathered on a daily basis and approved by the client as they are incurred
- make sure cost reports are prepared monthly, are accurate, and provide sufficient information for losses or profits to be tracked
- analyse the cost reports and take appropriate action to prevent further losses and where possible to recover the losses incurred
- check that materials are reconciled, and monitor losses so that appropriate action can be taken to prevent further losses
- undertake basic daily site costing to ensure the work is done in a cost efficient manner
- provide feedback to project staff of where the project is losing or making money
- ensure goods, services and expenses are collated, and paid correctly and on time
- check, approve, and submit personal expense claims as soon as possible with supporting documentation

Chapter 12 - Contractual

Changes, both minor and significant, will occur on every construction project. I've had projects whose final payment account has been double the original tender value. It is vital changes like these are tracked, and the client and their management team alerted to them as soon as the contractor becomes aware of them.

Changes are a normal part of the construction project process and will vary from project to project, both in the quantity, type and the quantum of the change. The important aspect of these changes is that they are handled in a timely, professional manner and in accordance with the conditions of the contract documentation for the project. Since contract documentation will vary from project to project, as will the client's expectations, it's important the Project Manager deals with these changes in an appropriate manner in terms of that project.

Some changes may result in no additional cost, or change in duration, while some may even result in a reduced cost or duration, however, they should all be documented and recorded.

All project staff should be aware of the specific procedures to follow when undertaking variation work.

Contractual issues are important. There are many valuable books and courses which cover this topic better than I can, and I would recommend that Project Managers attend specific contractual courses to obtain a better understanding.

Needless to say I've seen many serious and costly contractual mistakes made by Project Managers who were not as contractually astute as they should have been. Unfortunately I've also made some of these mistakes, even after being in the construction business for more than twenty-eight years.

Case study:
One of my projects was the civil works for a substation where we were a subcontractor to a main contractor. From the start the main contractor delayed us by issuing drawings and information late. In addition, we encountered extensive rock on the site which was excluded in our tender. Furthermore, the construction drawings differed from the tender drawings, and they were changed as work progressed resulting in us having to redo completed work. We documented all the delays and changes, and notified the main contractor via appropriate correspondence.

As happens with many projects, we were issued a letter of intent to start the project, but were only issued the contract document two months into the contract, which we proceeded to sign two weeks later. The contract document included the schedule which was agreed to at the start of the project before the delays were incurred.

Throughout the project the main contractor did not respond to any of our variations or claims for extension of time. When we finally got the contractor to sit around the table to resolve all the variations and extension of time claims we had lodged they refuted the extension of time claims on the basis that we had signed the contract document, which included the original approved schedule, after most of the delay events had occurred. By doing so, we had inadvertently accepted the original schedule as though it included the delays which occurred between project start and the date the contract was signed.

The lesson from the above is you should not start a contract before the contract document has been agreed and signed. If, for some unavoidable reason, this does not happen, then the Project Manager must ensure the contract eventually signed takes into account all changes, variations, delays and new information which occurred up to the date of signing. Alternatively the Project Manager could include a letter stating that the contract document signed only takes into account information that was known at the start of the project.

Furthermore, the contractor is at enormous risk working on a contract before the document is signed, with little protection if things go wrong. The contract document is there to protect both the client and the contractor, and I would not want to be work on a contract without one in place.

Many clients issue a letter of intent to proceed, but often these letters are of no use to the contractor when there are contractual problems on the project.

Also, once work has started on the contract the contractor is in a relatively weak position to negotiate any terms or conditions which the client may have introduced, and it's difficult to stop work once started as it can be costly and end up breaching the contract.

The best way to avoid disputes is for both parties to conduct the project in a manner of mutual trust, collaboration, and by acting in good faith. The emphasis is on both parties, but often one party fails to do this causing a dispute.

The Project Manager can help avoid disputes by understanding the contract, communicating effectively, putting in place clear processes with proper controls, and by educating themselves and their staff on contractual issues.

Of course, the best way of avoiding disputes is by not contracting on the wrong projects, or with the wrong client – something that Project Managers can't always control.

Contract documents

A contract is a legal binding document between parties. The parties are normally a client (with specific requirements) and the contractor whom they pay to carry out the works necessary to realise those requirements. Sometimes the client may be another contractor. The parties are legally obligated to comply and meet the conditions in the contract, and must agree to these terms and conditions, then sign and acknowledge them.

The document normally consists of the:
- formal instrument of agreement
- general conditions
- specific or special conditions
- specifications
- references to drawings
- addenda

The document must comply with the laws of the country and all parties must be legally compliant (if one of the parties is not legally compliant it could nullify the contract). The contract must be signed by representatives of the parties who have the necessary authority to sign these documents.

There are many different forms of contract used in different parts of the world and no Project Manager can be an expert on all of them, however, they should be aware of some common pitfalls. Many clients use a standard pro-forma for the general conditions of contract, and most of these documents say much the same thing, with specific clauses and conditions changed to suit the client. However, some of the contracts are altered in such a way that the contractor's rights are minimised which exposes them to an unfair risk.

The contract should spell out the rights and obligations of all the parties.

Contract documents should be kept as simple as possible. Often clauses conflict with one and another, or there is a conflict between the different drawings and specifications, so usually the contract specifies the precedence in which the documents must be read. If there's no set order specified there can be confusion.

There are various types of contracts according to how the contract is reimbursed, namely:
- lump-sum or fixed-price
- re-measurable or unit price
- cost-plus
- cost-plus to a maximum price
- target-plus

Contracts may also be:
- construction only
- design and construct
- design, build and operate
- design, build and finance
- design, build, operate and transfer

It's important to understand the particular contract conditions applicable to the project. I would recommend when starting a new project, always make a copy of the contract document, including all the special conditions, and the standard form of contract the document refers to, then highlight important paragraphs, and summarise salient points.

Items to note would be the payment terms and conditions, reasons for delays and variations, insurances, time for notifications, key dates, time for submissions of claims and variations, and who in the contract has the authority to issue variations.

If there is uncertainty regarding particular contractual issues request advice from senior managers, or people within the company who have contractual knowledge and experience. If advice is not available approach an external contractual practitioner which may save the project substantial sums of money, by helping to avoid contractual disputes at a later date.

Checking contract documents

When the document is issued for signing it must be reviewed, to check it's the same as the tender documentation, that is, unless changes have been mutually agreed. Ensure that:
- contract clauses are the same

- wording is unchanged
- specifications are the same
- no new drawings have been added
- drawing revisions are the same
- check that the contractor's exclusions in the tender have been included
- the schedule is correct
- the contract price is correct

I have often had contract documents issued to me that differed from the tender documents. Failure to pick these differences up can result in the contractor signing a document that contains items that were not originally priced and allowed for.

Laws governing the contract

The Project Manager must understand the laws governing the contract. These laws vary from country to country, and from state to state. Clients and managing contractors often operate in a different country, and elect to use the laws of their home country, so the contract may not necessarily be within the jurisdiction of the laws of the country or state where the project is located. These laws and legal systems may be unfamiliar to the contractor, and can leave the unsuspecting contractor exposed to working within a legal system they aren't familiar with.

The same contract – but is it?

Project Managers often become familiar with particular contract conditions, or repeatedly work for the same client using the same contract conditions. It is, however, important to read each contract carefully, because contract conditions can often appear the same, but there may be subtle rewording of clauses which can impact on the way the contract is read and administered. Never assume the current contract is the same as the previous one.

Letters of intent

Sometimes a client requires the contractor to start the project as soon as possible, and rather than delaying the project while the contract document is prepared, the client will issue a letter of intent. Often letters of intent are very loose documents, a few lines in length, which can be dangerous for the contractor, offering little protection if the project is terminated.

If the contractor is prepared to work under a letter of intent while the formal contract is prepared it's important the Project Manager carefully checks to ensure it:
- clearly defines the scope of the works
- includes the basis of payment
- states the value of the work to be completed
- has the start and completion dates of the project
- references the tender, including the contractor's clarifications
- has a termination clause
- includes a date by which the contract document will be issued

The Project Manager must constantly check the value of work done, and ensure the amount in the letter of intent is not exceeded. If it seems the value will exceed the that stated, the Project Manager must timeously warn the client so a revised letter can be issued, because if it is exceeded the contractor faces the risk of not being paid for this work.

Documentation and records

As mentioned previously, correct and accurate documentation is vital on any project, particularly should a dispute arises, or when submitting a variation. Documentation which can become important includes:

- photographs showing progress, variations and completed work
- minutes of meetings, with the client and subcontractor
- daily diaries
- correspondence received and sent
- information requests
- drawing registers
- drawing issue receipts
- site instructions
- notifications
- the approved contract schedule
- progress updates
- the signed contract document
- the tender (tender submission, tender correspondence, tender schedule, tender drawings and specifications and tender meeting minutes)
- signed day-works sheets

Client's obligations

The client has certain responsibilities they need to comply with to ensure the contractor can carry out the project works (these are normally outlined within the contract documents). If they aren't the contractor should clarify them during the tender stage. The client's responsibilities include, amongst others:

- ensuring they have the project finance in place to enable payment of the contractor
- pay all parties on time for services rendered in accordance with the contract
- demarcating property boundaries
- ensuring they have ownership of the property
- putting all approvals in place, such as planning, building, traffic, and environmental
- appointing an appropriately qualified design and works supervision team
- if there is no managing contractor they should manage the design team, including but not limited to ensuring the team produces appropriate information on time, to the correct quality requirements, and in accordance with the project scope, and that the team is responsive to problems and queries
- ensuring there is cooperation amongst all members of the contract team
- ensuring project issues are promptly and affectively dealt with

The Project Manager should ensure that the client complies with these responsibilities since often the client is inexperienced, or stands back from the project, which can result in the contractor having to undertake responsibilities for which they have not allowed. Unfortunately, sometimes the contractor has to manage the client to ensure the project is a success.

Designer's obligations

The designer must ensure:
- their designs comply with local, state and country design codes
- they supply the appropriate setting out data
- they produce all drawings and specifications required for the contractor to construct the project
- their designs are accurate and workable
- they take responsibility for the integrity of the design
- the design conforms to the client's scope
- they have obtained the necessary approvals for the design
- they respond promptly to queries from the contractor
- they provide prompt feedback to the contractor's designs, shop drawings and samples, and not unnecessarily withhold these approvals
- they act in a fair and impartial way
- if it's their responsibility to evaluate the contractor's work that this is done in such a way that their design is not compromised, is done on a regular ongoing basis, and the contractor is informed immediately if any deviations are detected

Unfortunately not all of these responsibilities are usually clear in the document, and many design teams neglect to perform all of their obligations as they should, which can lead to information of a poor quality being issued late.

Managing contractor's obligations

On many projects the client will appoint a managing contractor to manage some of their responsibilities. It's important, though, that the client understands that this does not absolve them of their responsibilities.

The managing contractor must:
- manage and coordinate the various contractors
- manage the design team
- protect the client
- ensure the project is delivered on time, to the required quality and specifications, meeting the client's brief, without accidents and within budget
- depending on the contract, ensure that the various services are available for the contractor
- ensure the contractor has the required access to carry out their works
- ensure the client pays the various parties fairly for services rendered in accordance with their contracts
- ensure the information supplied by the design team is provided according to schedule and the information is of sufficient quality to enable the construction to proceed

Contractor's obligations

The contractor has certain responsibilities and it's important the Project Manager understands what these are. These include:
- managing and supervising the project to ensure it is successful and:
    - is completed on time
    - meets the quality standards
    - complies with the specifications
    - has no safety or environmental incidents
    - has no industrial relations issues
- having a duty to enquire (this would include querying discrepancies in information and informing the client of any concerns they have with the design or construction methods)
- reviewing drawings issued for construction
- planning and scheduling the works
- coordinating the works
- setting-out the works
- notifying the client timeously of unexpected problems and variations
- invoicing for the work completed
- ensuring that all the required insurances are in place
- complying with the terms and conditions of the contract including submitting all contractual deliverables

Types of claims

There are many different reasons for a claim and the Project Manager must evaluate these to ensure that each is valid within the terms of the project's contract document, since a claim valid in terms of one contract may not be valid under another. Claims may be for any of the following reasons:
- errors and omissions in the document
- changes to specifications
- changes to drawings
- unforseen conditions, such as:
    - unforseen ground conditions
    - restricted access
    - discovery of artefacts
    - uncovering of dangerous chemicals or hazardous materials
    - the unexpected presence of underground water
- changes of conditions
- clients accelerating the schedule, delaying or changing completion dates
- clients not providing access on time
- drawings issued late
- clients not providing services or failing to provide them on time
- client supplied equipment or materials arriving late, or not to the correct quality
- impact caused by the client's direct contractors on the project work area such as closing off access routes
- changes to scope

- drawing errors and drawing coordination problems
- changes of law within the state or country
- unseasonable adverse weather

The client normally has the right to vary the project providing the variation is reasonable, and the Project Manager can't normally refuse to accelerate, undertake additional work and alter the schedule to meet revised completion or access date. However, the contractor has the right to claim the additional costs incurred to execute the variation.

It's important the appropriate person within the client's organisation signs the variation. I have on occasions had client's representatives, without the correct authority, sign and issue a variation, thereby making it invalid.

Variations should be logged in a register since often the contractor forgets about it and it's not executed, which can delay the final completion of the project. The Project Manager should track the variation register to ensure that variations are responded to by both the client and contractor within the specified time-frames. The variation register should be presented as part of the project progress meetings and in the monthly project reports.

Informing the client

Part of avoiding disputes is ensuring the client is kept informed of all possible delays, claims and variations. Preferably this should occur before the event, or as soon as possible after the event has occurred, and certainly within the time period specified in the contract for notifications.

Drafting a claim

Claims should have, at a minimum:
- a description of the event
- the cause of the event
- the date of the event
- the impact of the event
- steps taken to mitigate the impact
- the cost and time impacts of the event
- all supporting documentation attached, or should refer to supporting documentation referencing:
    - clause numbers
    - drawing numbers
    - schedule item numbers
    - correspondence
    - specifications

It's essential that this supporting documentation is relevant to the claim, supports the claim and is not contradictory (any contradictions must be explained as part of the claim).

As part of formulating the impact of the event all calculations and schedules should be included. The claim schedule should reference the approved contract schedule. Calculations should reference where the facts and figures came from and how they were put together. The calculations should be checked for arithmetical errors (which can occur all too frequently).

The claim must:
- be lodged within the time-frame specified in the contract
- be addressed to the correct person
- be delivered to the correct address

Acceleration

Acceleration may be required if:
- the scope of works has increased, but the completion date remains the same
- the contractor has suffered excusable delays entitling them to an extension of time, but the completion date remains unchanged
- the client requires an earlier completion date to the one in the contract

Normally I wouldn't accelerate the works without the client issuing an instruction to do so, and preferably only after the costs of it have been agreed and the acceleration schedule has been accepted as the new contract schedule. In most cases, however, the client is reluctant to issue an instruction to accelerate, despite delays attributable to the client, or the addition of work. Yet, at the same time they may refuse to vary the completion date, often hiding behind a clause in the contract stating that the contractor must take mitigating steps to avoid and recover any delays, refusing then, to accept any schedule which doesn't comply with their milestones. Project Managers must resist the urge to defer to clients in this regard, or they will find themselves accelerating the works without being compensated for their additional costs

In this case, it's important the contractor submits a fully substantiated claim for an extension of time, after which the contractor may then action a constructive acceleration to meet the client's completion dates. The Project Manager should formally notify the client that the contractor is accelerating to meet the completion date, and that the contractor expects to be reimbursed the costs incurred for accelerating.

There is some danger that the contractor will not be paid their acceleration costs if they accelerate, and still fail to meet the required completion dates.

Force majeure

This is a term given to delays which none of the contracting parties control, or is responsible for, such as unusual adverse weather conditions and national strikes. Normally the contract document makes reference to force majeure, and the contractor may be entitled to an extension of time covering the delays incurred due to the event, however, in most cases each party bears their own costs.

The Project Manager must be aware of which events are force majeure, and which are not, as it's often more beneficial to claim a delay for another reason and under another clause.

Concurrency of delays

Some delays and events occur concurrently, and sometimes these are a result of faults by both the client and the contractor, and occasionally even a force majeure event. It's often difficult to resolve who is responsible for the overall delay and quantify it. It may, therefore, be necessary to put the individual delays together to see their overall effect.

When considering the cause of the delay, the norm is to attribute the cause to whoever caused the first delay, with the second delay only coming into effect once the first delay is over. For example, if the contractor delayed themselves by three days due to material arriving late, and on the second day the client delayed the contractor by ten days because a drawing was issued late, the delay the contractor could claim would be ten days minus two days, (the contractor had caused themselves the first delay of three days and one of these had elapsed when the client caused a delay, in other words the delay due to lack of materials overlapped the client's delay by two days), equalling eight days. However, if the client first caused the delay of ten days due to late information, and three days after the client's delay the contractor delayed themselves by three days due to materials arriving late, then the delay caused by the client would be ten days. Of course most contractors will in both cases always try and claim the ten days delay caused by the client.

Obviously this would vary from contract to contract, and in many instances some common sense, and give-and-take, should prevail from all parties.

Site instructions

The Project Manager should ensure their staff understands the importance of not accepting verbal instructions. I've often witnessed the managing contractor requesting Supervisors do additional work, or do work differently to that shown on drawings. But if this request is not in writing there is a risk the contractor will not be paid for the additional work, and further more if structures are not constructed according to the drawings they may have to be rebuilt.

Changes to the scope of work should only be issued in the form of a site instruction signed by the client's authorised person. Sometimes a person may only be authorised to issue an instruction for a variation providing there's no cost, or it doesn't exceed a prescribed monetary limit. So the Project Manager must ensure they, and the construction team, understand these limitations of authority.

Equally important is to ensure that only members of the contractor's staff authorised to receive site instructions sign and receive them.

The person receiving the instruction should carefully read the wording on the instruction, because often the client has included wording to the effect that no cost or additional time is granted for the works, sometimes it's just a standard printed statement on the client's site instruction form. Either way, this statement should be deleted, because the contractor receiving the instruction is usually not in a position to quantify if there will be time or cost impacts. Also check the wording for other potential problems, for example, the instruction could be worded in a way that it appears the work is being redone due to the contractor's defective workmanship, or there may be a completion date which is not achievable, or it could stipulate the works will be paid for at standard contract rates which might not be reasonable if the works are small or access is limited.

Many clients' representatives are reluctant to issue site instructions, yet they will happily issue verbal instructions or send out emails. It's the Project Manager's job to ensure the client issues a formal site instruction since this will formalise the payment process and also prevent miscommunications. Often we have been cooperative, and done work the client verbally requested, without waiting for a formal site instruction, yet not been paid because the client's representative later denied making the request, or claimed we executed the work incorrectly.

Complying with a site instruction and carrying out the work as directed doesn't necessarily mean you will be paid it. There is still a process to be followed to ensure the payment. This process may entail first pricing the instruction, and then a variation may have to be issued by the client. The Project Manager must ensure all staff are aware of these procedures, because I've had clients claim that if they had known the cost of the work before it was executed, they wouldn't have requested it to go ahead.

Insurance claims

When an insurable event occurs it's important that the relevant insurance company is notified as soon as possible. The Project Manager must know if the event is insurable, under what policy it can be claimed, who the insurer is, and what the excess is. Photographs should be taken of the damage, and the area must be made safe and secured. A detailed report should also be prepared setting out the events leading up to the damage and an estimate of the repair costs.

When the insurer is contacted they will normally advise if an assessor will inspect the damage and what the next steps are to repair it. Then, once the assessor gives the go ahead to proceed the damage can be repaired.

Obviously time is of the essence since the damage may be delaying progress on the project, and normally most insurances will not cover the impact of these delays.

Resolving disputes

There are various options to resolve disputes.
- The cheapest and easiest is to negotiate with the individuals directly involved with the project, which may require some compromise from both parties.
- If negotiation fails the dispute should be referred to senior managers from the different parties. Often the problem is due to a clash of personality, or an individual's incorrect interpretation of the contract, and most managers can quickly resolve the issue when these obstacles are removed.
- If the owner of the facility is not party to the dispute they can be approached to assist in resolving the dispute.

The contract document often dictates the dispute resolution process to be followed should negotiation fail. Normally this route includes one or more of the following:
- Mediation, where an independent mediator is appointed to get the parties together to discuss and resolve the problem.
- Arbitration, where an independent arbitrator is appointed to hear evidence and then make a ruling on what they believe is the correct solution. Arbitration is usually not binding.
- Litigation tends to take a more legal approach with the process focusing on the legal rights of a party, without necessarily understanding the project and the impact of variations on the construction processes and schedule.

Arbitration and litigation can be drawn out processes, often taking several years to resolve, they are also costly, and may not provide the answer either party was expecting.

Some projects require a dispute resolution board to be appointed at the start of the project which is usually made up of three members (one chosen by the contractor, another by the client, and the third by the first two members). Both parties are usually responsible for the costs of the board. The board usually visits the project monthly and becomes aware of issues and problems as they arise. Should a dispute arise it's passed to the board for a ruling.

Depending on the country there may be other options available to contractors, so it's worthwhile consulting experts.

Notices

Notices issued to the client or a subcontractor must:
- be dated
- be addressed to the correct person (usually nominated in the contract) at the correct address
- be delivered in accordance with the contract (some notices must be hand delivered to a physical address and signed for, and even if there is no provision in the contract for an acknowledgement to be signed I would urge that receipt of the notice should be acknowledged and signed)
- refer to the correct clause and reason for the notice
- refer to steps already taken leading up to the issuing of the notice
- normally demand an action by a particular date
- outline the consequences of failing to implement the actions by this date
- be signed with the name and contact details of the person issuing the notice
- give the details where the reply should be addressed to

Liens

Some contracts allow the contractor to impose a lien on the project in the event of a dispute or for non-payment. Legal advice should be sought before exercising a lien on the facility to ensure the client has received the correct notices and the lien complies with the contract.

Contractors should also be aware that subcontractors may have a right to take out a lien on the facility which may prevent the contractor from handing the project over to the client. The Project Manager should therefore ensure payment or contractual issues with a subcontractor are resolved as quickly as possible so they do not disrupt the project.

Terminating the contract

Both the contractor and the client are able to terminate the contract under certain circumstances and situations. There is normally a termination clause in the contract. Termination should be used as a last resort since it can be a costly process for all parties. It's important when terminating the contract the Project Manager does it strictly in accordance with the contract, since termination for the wrong reason, or not following the termination process correctly, could result in the contractor being in breach of the contract.

Liquidated damages

The Project Manager must understand, and be aware of when the client can enforce liquidated damages. Damages can normally be imposed when a contractual date has not been met, or the specific contractual obligations haven't been fulfilled. Liquidated damages should only be applied by the client if they have been physically delayed from taking occupation. In the past I've had clients wrongfully impose damages on us after we missed a milestone, even though they weren't ready to occupy the section. Liquidated damages shouldn't be applied for minor incomplete works which don't prevent the client from taking occupation of the facilities.

Summary

- The contract protects the client and the contractor, and spells out their rights and obligations.
- The contract should be signed by both parties before work starts.
- The contract should be carefully compared to the tender documentation before it's signed to ensure the conditions are the same as those tendered for.
- A letter of intent should refer to the tender submission and conditions, and the Project Manager must ensure their rights are protected, and that work doesn't exceed the value stated in the letter.
- The Project Manager must understand the contract documentation to ensure their compliance with the contract.
- It's essential that the Project Manager communicates with the client and that all documentation is kept current and accurate.
- Variations must be addressed correctly and must include all supporting documentation and calculations, or should correctly reference these.
- There are different dispute resolution procedures with many contracts outlining the procedures to follow.
- The Project Manager should avoid disputes, or if possible resolve them without going to a court of law since this can be costly and time consuming.
- If a contract is terminated due to a breach, notices must be filed correctly and in terms of the contract.
- Site instructions and variations must be in writing and should be checked to ensure they are signed by an authorised person and that the wording is acceptable.
- Only the contractor's authorised persons should accept site instructions or variations.
- The contractor must be aware of the milestones with liquidated damages attached, and understand the obligations to be fulfilled to meet them.

Chapter 13 - Completing and Closing Out the Project

Case study:
A few months after completing a project, and demobilising from site, the client called saying there was a concrete footing in our scope that we hadn't built. After checking the drawings I concluded we had mistakenly left the footing out despite it being clearly shown on drawings from the beginning.

With no other choice, we assembled a team from our nearest project, three hundred kilometres away, and returned to construct the missing foundation, which required only half a cubic metre of concrete with minor excavations and formwork. Including travelling the work took three people three days, compared to one person for half a day if the work had been completed in the course of the project. In addition we had to pay a premium for the small amount of concrete required.

In all, the exercise probably cost the company a hundred fold more than it would have if we had constructed the footing while still on site.

The lesson here is to ensure all the work is complete before demobilising from the project. There are many projects we've been summoned back to after demobilising, so as to resolve problems, many of which should have been fixed before leaving the project.

When nearing the end of the project it is good practice to prepare a finishing schedule. This should include all the remaining items of work and time for completing punch list items.

Closing out a project can often be a complicated and time-consuming process that is underestimated and overlooked by Project Managers and staff who may already be anticipating moving to their next projects. If proper planning and preparation is done at an early stage the close out process will be much simpler.

Keeping documentation up to date, and filed in a correct and orderly system will ensure the documentation is substantially complete, and only requires minimal work before it's handed over to the client. However, I've rarely found it done this way, resulting in some project staff working for months after the completion date, trying to recover data and put it into some semblance of order to hand over to the client. This is costly, delays personnel from moving to another project, and holds up final payments and the release of guarantees.

I also suggest preparing a list of items the contractor must supply to the client as part of the handover process. Delegate someone to be responsible for each item, and a date by when it should be closed-out. These items are often listed in the contract document, and would include quality control documentation (including closing out of all punch list items and non-conformance reports), a list of spares, spare parts, additional (or attic) stock, maintenance and operations manuals, warranties and guarantees, commissioning data, and as-built drawings.

Chapter 13 - Completing and Closing Out the Project

Stages of completion

There are various stages of completion, so the Project Manager must read the contract document to understand what these are, and what is required to meet them. (It should be noted that terminologies vary between different contracts so it's important Project Managers read and understand the definitions in their particular contract document.)

Substantial completion should normally be issued when the facility is acceptable for the owner to occupy and use, when only minor punch list items are remaining, which do not affect the client from safely using the facility for its intended purpose. The certificate of substantial completion would include a list of the outstanding items. The completion date is important because:
- after this date liquidated damages cannot usually be applied
- in most cases it is when the warranty and defects liability period starts
- it's often the trigger to release the retention money or bond

A project may have multiple substantial completions for different sections or separable portions of the project, with one final completion when all the interim portions have been achieved.

Often a completion certificate requires the completion of certain tests, provision of all quality documentation, as well as commissioning and operational data. The client may also need to prepare a punch list. All of these tasks may take several weeks, so it's important to be prepared well before the time so as not to delay the issuing of the certificate. In addition, the client often requires timeous written notification.

In preparation for handover of the project, it's good practice to go through the process with the client to ensure it goes smoothly. Often the client is unprepared, or doesn't understand what is involved. On most projects when substantial completion is achieved the client has a number of duties and obligations which can include, ensuring they have sufficient insurance in place, they are able to take over the operations of the facility (if necessary including having the trained staff), and they take responsibility for its security.

Beneficial completion or practical completion is when a section of works is handed over to enable the client, or their contractors, to occupy the section and install further work. Before another contractor moves into a section of works a punch list should be prepared with the client to enable the recording of damages which are as a result of the client or their contractors.

At the end of the maintenance or defects liability period (which could be three, twelve or twenty-four months after substantial completion), the contractor would normally request the client draw up a final punch list. After this list has been attended to, the client should issue a certificate of final completion and release all retentions and guarantees.

Completing punch lists

The contractor should prepare punch lists for each structure as it's completed, then attend to and close them out as soon as possible. After completing these items the client should be requested, in writing, to draw up their official punch list, which should be formally issued to the contractor.

The Project Manager should get the various disciplines to prepare their punch lists and issue them together, that way all the trades can be attended to at the same time, minimising rework and cleaning.

(Refer to Chapter 9)

Contractors need to be aware that once sections of the project are handed over and the client takes occupation it's usually difficult for the contractor to gain access to complete punch list items. These often have to be done at times which suit the client, which may even include doing them after-hours, adding to the costs of attending to them.

Quality control and other documentation

Most projects require a full set of quality control documents to be handed to the client on completion of the project. Before handing these documents over ensure the file is in accordance with the client's requirements and includes:
- check lists
- test results
- as-built drawings
- operating and maintenance manuals
- closed out, non-conformances reports
- signed off punch lists

Ensure a copy of all the documentation is retained, and request the client to sign an acknowledgement of receipt, since I have had occasions where these documents were lost.

Certificate of occupancy

Some projects require that the contractor obtain the certificate of occupancy from local authorities and fire departments. Process facilities may require special licenses and permits to operate. So, the Project Manager must understand the relevant procedures to be followed, since some of them may take several weeks to obtain, and certain testing must have been completed.

As-built drawings

Most contracts require the contractor to supply as-built drawings. These should be prepared as the project progresses since some may be for services or structures that get covered up, buried or concealed. On a major project preparation can be a substantial undertaking and consideration may need to be given to having a staff member dedicated to this task. The drawing formats and presentation must comply with the client's requirements. Some clients may be happy with simple red line mark-ups of construction drawings, while other clients require electronic copies of the as-built drawings.

As with other documentation handed to the client, ensure copies of the drawings are kept, and that the client's representative signs a receipt for their drawings, since it'll be a costly exercise if they're lost and have to be redrawn.

Warranties and guarantees

The contract normally specifies the warranties required, although usually all items of equipment built into the project should have a warranty. These should be handed to the client on completion of the project.

A check list of the warranties should be drawn up as the project progresses, and be updated as each new item of equipment is ordered. A member of staff should be tasked with ensuring the warranties are received from the suppliers and subcontractors, that they are in compliance with the contract requirements, are valid for the period specified in the contract documents, and they should also specify the process to be followed and who to contact if the item breaks down. Usually the warranty is voided if a non-authorised workman repairs the item.

As the warranties are received they should be filed in the warranty folder for that section of the facility.

Tying into existing services

Many projects require new services and equipment to connect into existing services and systems. This can be a simple task, but in some cases a major undertaking when the tie-in affects the operations of the existing facilities, or other third party used or owned facilities, causing disruption.

There are a variety of steps that can be taken to increase the chances of this running smoothly:

- Prepare a plan covering the tie-in which will include:
    - the services to be connected
    - existing services which will be affected, such as existing electrical or water supply
    - the expected duration of the works, include a schedule if necessary
    - how the task will be carried out, including a method statement and the resources required
    - safety aspects relating to the new works and the existing works, since many of the tie-ins entail working in and around existing operational facilities which have additional risks, such as those to members of the public or other contractors
- Once the plan has been drawn-up, organise a meeting with all the affected parties and those that will be part of the tie-in process. Discuss and modify the plan, then when all parties are in agreement, a date and time can be proposed for the works. Sometimes it may only be possible to do the works after-hours.
- To undertake the works it may be necessary to obtain permission from traffic or local authorities and arrange permits.
- When all permits and permissions are in place, and the date and the time is confirmed, all parties and stakeholders must be given adequate notice of the event. This notification would include:
    - what services will be affected
    - the duration for the works
    - contingency plans in place during this period
    - the risks and dangers
- It's important the contractor is fully prepared to carry out the works since

there is little worse than cancelling a tie-in because the contractor was not adequately prepared. Preparation includes:
    - having adequate skilled workers and supervisors
    - briefing them on the task and how it will be done
    - having the correct equipment and scaffolding
    - back-up plans and equipment should be in place
    - on occasion senior members of the contractor's staff and even the Project Manager to be present
- After the tie-in is complete all parties and stakeholders should be notified of the successful (or otherwise) completion of the tie-in so they are aware that normal processes can resume.
- If the tie-in involves modifying an existing process the stakeholders must understand what has changed, and how it effects the operations of the existing facility, particularly relating to operation and maintenance.

Commissioning

Sometimes commissioning is required on projects, this could be as simple as plugging in a fridge and checking it works, or running and balancing an air-conditioning system, or as intricate as running and operating a new process facility ensuring that it meets the client's requirements and outputs. Some of these processes can be complex, and may even be unable to be completed until the client has installed all their services, and is ready to supply the feedstock and accept the product. In such cases, it will be necessary to prepare a proper commissioning plan and schedule, which allocates responsibilities for completing components, as well as their completion dates.

Good communication is the key to successful commissioning. It can be a dangerous process. As systems are brought on-line it's essential all parties working in the area are advised of the process. It may even be necessary to have exclusion zones both during the process and once the system is running. The commissioning plan must detail all safety concerns, procedures and contingency plans.

Commissioning staff must be trained and have access to all the equipment data and commissioning sheets because incorrect commissioning can damage plant and equipment.

Case study:
I heard of one instance when a major electrical substation was severely damaged during the commissioning of a copper process facility. The replacement components took over three months to obtain, resulting in the whole project being completed over three months late.

It's often a requirement for the client to witness the commissioning. Even if it's not, it's good practice to ensure they have an appropriate and knowledgeable person present who can verify everything is working.

Once the facility is commissioned it must be handed over to trained operators, and have all the materials, fuels and chemicals required to operate it.

Operation and maintenance, and training the client's staff

Many contracts require the contractor to operate and maintain equipment for a period of time. Sometimes this is only a short period to prove to the client the equipment performs according to the contract design specifications. Occasionally though when handing over sections of the project there may be a requirement to have a section operating to enable the balance of the project to be completed. For an operations and maintenance contract the contractor may have to operate the facility for a period of several months, or years, after it's built.

If the operation of equipment is in the project scope the Project Manager must ensure there are:

- suitable, trained and qualified personnel (if it's a 24-hour operation, a night shift may be required as well)
- the correct spare parts readily available in case there's a failure of the equipment
- sufficient stocks of the chemicals required for the processes to operate (this particularly applies to water and sewer treatment plants)

The Project Manager should check the contract timeously to understand what is in the contractor's scope and what the client must supply.

Most contracts require the contractor supply operating and maintenance manuals for all equipment installed by the contractor. These documents would contain:

- relevant information about the systems
- installation instructions
- safe operating instructions
- maintenance instructions
- details of the warranty
- contact details of the supplier
- a list of spare parts

It's important that these manuals are collected as the equipment arrives throughout the project, and that they are filed in a folder in the relevant section of work.

Spare parts

Ensure the required spare parts are handed over to the client's representative. They should:

- be signed for (I have experienced spare parts go missing or get damaged after they were handed over)
- be clearly marked with the project name as well as the item the part is for
- be properly packaged and correctly stored, always ensuring the client's representative understands any special storage instructions (put these in writing if necessary)

At the start of the project, prepare a schedule with all the spare parts required for each item of equipment, and monitor the list to track who the supplier is, when they are expected, which parts have arrived and which have been handed over to the client. The contract document would normally specify what spares are required and the list should include spares that subcontractors will supply as part of their contract.

These parts should be ordered at the time of placing the order for the equipment since many can take some time to procure.

Plant and equipment

All hired plant and equipment must be returned to suppliers so unnecessary costs are not incurred. At the start of the contract a schedule of hired plant and equipment should be prepared and as new items are rented these should be added, then as the equipment is returned it should be removed from the list. If this has been done it will be easy to see any outstanding items. Remember all items must be sent back to the supplier and not just put off hire.

Often when a Supervisor is transferred to another project he inadvertently takes a rented piece of equipment with his tools. In which case several months after the project has been completed the hirer is still invoicing for the item. It then takes a lot of effort to trace the item since project staff have moved to other contracts and the project documentation has been archived, so avoid this if possible.

In some cases, the item may have been lost or stolen, in which case the Project Manager must arrange for the supplier to provide a cost to purchase the item, so it can be purchased and the hire charges stopped.

If there's no list of hired equipment, or if there are doubts about the accuracy of this list, it's good practice to contact the hire companies and request a list of outstanding items on hire. I've found in the past that hire companies can several months after the project has closed, produce a long list of outstanding equipment. Sometimes I believe this list may be fictitious, due to the assumption that after so long many of the staff who worked on the project will be uncontactable and the paperwork no longer be available.

Financial

It's frustrating when months, even years after a project has been completed, suppliers are sending queries regarding payments for items they provided during the construction period. Often by this stage the project staff have moved onto other projects, some may no longer be in the employ of the company, and the relevant documentation may be archived or lost.

Therefore, it's good practice at the end of the project to approach all major suppliers and contractors to check there's no outstanding money or invoices. All queries and disputes should be resolved before the project is closed out.

Closing out subcontractors

Subcontractors must complete all punch lists, hand over all quality documentation, as-built drawings, guarantees, warranties, spares, operating manuals, and commissioning data. The final account must be agreed, including any deductions made to their account.

An important aspect of closing out subcontractors on a project is to review their performance. This can be done using a standard form, and should evaluate the subcontractor's safety, quality and schedule performance, as well as the cooperation received from them during the project. The review report should be distributed to the estimating department and other Project Managers.

Chapter 13 - Completing and Closing Out the Project

Demobilisation of staff, personnel and equipment
- Notify suppliers their equipment is off hire and arrange transport to remove the equipment and temporary facilities.
- Ensure the correct job hazard assessments are in place to cover the loading of the equipment.
- Items of equipment purchased specifically for the project can either be sold or transferred to another project.
- Excess material may be able to be returned to the suppliers, however often the transport costs are more than the credit the supplier provides. Alternatively, the materials may be sold to neighbours providing they take responsibility for them.
- Excess materials that the client has paid for must be handed over to the client, and these should be signed for and stacked in their designated area.
- Prepare a list of personnel that will be demobilised from the project with their expected demobilisation dates, trades, and the conditions under which they were employed (some workers may be permanent employees while others may have been employed specifically for the project). Provide the list to the human resources department, and to other Project Managers who may require personnel.
- Should there be no further positions available, some personnel may have to be terminated. Obtain advice on this process and ensure that it's followed correctly and personnel are given the required notifications. Then gather all their documentation together, and ensure they receive the correct pay and documentation when they leave.

Clearing laydown areas

Laydown and site office areas must be cleared at the end of the project. This will entail removing footings, cables, underground piping, waste and excess material. If the client or another subcontractor is happy for foundations or materials to remain, obtain a letter from them accepting responsibility for their removal. Alternatively, some projects may require the laydown areas and access roads to be rehabilitated which may entail shaping, topsoiling and revegetating them with local vegetation to ensure the areas will not be eroded. These areas should be inspected and signed off by the client as being acceptable, then take photographs to verify everything has been cleaned up.

Termination of services and accommodation

At the end of the project accommodation must be handed back to the client or external providers. The accommodation should be inspected with the owners, or their representatives, to ensure it hasn't been damaged. At this point there are normally bonds and deposits, which could be of significant value, to be recovered from the owners.

All services must be terminated both to the accommodation and site offices, requiring the necessary authorities be advised. Often there are deposits here as well that should be tracked to ensure they are refunded.

Final account

In most contracts the final statement of account must be prepared within a specified time after project completion. The team working on the final valuation and claims, should be aware of this date, and may require a schedule to ensure all the work is completed within this time. It's in the interest of all parties that the final account is resolved as soon as possible while most of the client's and contractor's personnel are still available, the project is fresh in their minds, and information is still readily available.

On conclusion of the final account it should be signed by all parties acknowledging that there are no further costs or claims from any of the parties.

Return of bonds and sureties

The Project Manager should check when the sureties and bonds can be released, and ensure all paperwork is in place and requirements met, then submit a formal letter requesting their release. Sometimes these bonds are only released at the end of the maintenance period and even though the Project Manager may be working on another project it's important they ensure the bond is returned as soon as possible, since bonds that aren't returned will attract additional costs from the bank or insurance company that issued them.

Archiving records

The project records must be archived in a safe place where they will not be damaged and can be easily retrieved. Records normally have to be kept for five or more years, but this can vary between clients, different states and countries. I would actually suggest they are saved for a longer period, since some of my projects have seen us go back several years after completion, and referring to records saved us considerable costs.

There are also electronic forms of storage which can make retrieval easier. Some of these are web-based allowing the contractor, client and other involved parties to access the documents.

Lessons learned

It's good practice to arrange 'a lessons learned' workshop at the end of every project. I've seen many mistakes made, and then the company, and even the same team on occasion, go and make the same mistakes on their next project. The most important part of making a mistake is being able to fix it. The second most important part is to learn from the mistake and avoid making it again. We should always pass our knowledge on to other people within the company so they can also learn from our mistakes.

Of course, it's not only about learning from mistakes. When we manage projects successfully it's just as important to pass the lessons of these successes on to other people.

I would suggest organising an internal workshop with key members of the project team, including safety, quality, engineering, subcontractor management, administration, planning, and key Supervisors. Of course, you may have run a small project doing most of the things yourself, with the help of one or two Supervisors, but it's just as important to sit down with a planned agenda and workshop the lessons learned on the project.

The first part of the exercise should be to go through each of the disciplines on the project and work out what was executed well and a success, and what was done badly resulting in a problem. Do the same exercises then with the company Head Office departments to see where they helped and where they hindered, and then repeat this for the client and their management team.

After listing the successes and failures on the project analyse them to understand the causes. The next step is to work out what should have, and could have, been done differently. Sometimes of course, you may find there was nothing that could have been done differently, although this is seldom the case.

After completing this exercise internally, the information needs to be passed on to those in the company who may benefit from hearing the results. It's important that this exercise should not be seen as a blame game, a 'get out of gaol card' for the project team, nor a finger pointing exercise. The objective is to improve processes within the project team, within the company, and most importantly to learn from past experiences and become more successful.

Even if you're unable to get any departments within the company to change, you should at least have learned where there are problems within the company structure, where you can rely on support, and where there may be shortcomings.

Lessons, both good and bad, can be learned about specific products or subcontractors and these should be passed on to other Project Managers within the company in the form of an emailed memo, or even by conducting a short presentation attended by the Project Managers and others involved.

Summary

- The conclusion of a project must be done in an orderly fashion, with a completion lists and schedule prepared to ensure all items are attended to.
- The required documentation must be prepared and handed over to the client; this includes quality documentation, test results, as-built drawings, spares list, operations and maintenance manuals, guarantees, warranties and commissioning data.
- If required, the client's personnel must be trained in the operation of the facility.
- Punch lists must be drawn up and attended to by the contractor and subcontractors.
- Commissioning is an important process that must be planned, coordinated, and carried out by competent knowledgeable staff, because improper commissioning can lead to accidents or damage to property.
- Personnel and equipment must be demobilised, services terminated and laydown areas cleaned and handed back.
- Deposits paid for services and accommodation should be recovered.
- All outstanding financial issues with suppliers and subcontractors must be resolved, and the final account agreed and signed with the client.
- All project documentation must be correctly filed and sent to safe but accessible storage.

Conclusion

By now you may be wondering if I made money on any of the projects I've been involved with, especially since some appear to be one long list of problems and mistakes. The truth is most of my projects were successful, some of this may be ascribed to luck, although it has been said you make your own luck. We, however, should have done things better on many occasions, and could have been even more financially successful.

I hope that you've learned from some of my experiences and will not make the same mistakes my team and I made.

In mentioning my team I do have to say that, in general, I always worked with a great team, all willing to learn, and I certainly wouldn't have reached the positions I did without them.

As I said at the beginning, this is not a technical book, and nor is it a book that will teach you how to become a Project Manager. It's not going to teach you how to build great structures like the Sydney Harbour Bridge, the Golden Gate Bridge or the Olympic Stadium in Beijing, but it is a practical guide on how to manage your construction projects.

You may be asking if you have to do everything mentioned in the book. The short answer is probably yes, but, it will vary from project to project. You may be fortunate and be able to delegate some of the tasks to members of your staff, but even then, you'll need to follow up to check they are being carried out correctly. Indeed, managing a construction project is not an easy job. Many Engineers and Supervisors think it's only a short step from where they are. Let me assure you, this is not usually the case.

A Project Manager has an enormous responsibility. The very lives of the workers depend on you and the way you manage and plan the project.

Seek help from experienced and knowledgeable sources where possible. Ask questions. Learn from your mistakes. Have a positive learning attitude. Be honest and fair. Help, coach and mentor others. But most importantly have fun, enjoy what you're doing and help others to enjoy working with you.

Glossary

Terminologies vary between different construction industries, countries and even companies. The descriptions below relate more to their meaning within the book and aren't necessarily their official descriptions.

Acceleration – to shorten the schedule, or program, so the project is completed earlier, or alternatively, to complete more work in the same time period.
Activity – an individual task or event on the schedule.
Allowable – the estimated cost for a particular activity, or task, allowed for in the tender or project budget.
As-built drawings – drawings that are prepared by the contractor to show the position and final dimensions of the structure as constructed.
Back-charges – money charged to a subcontractor for costs the contractor incurred to carry out or rectify the subcontractor's work.
Bonds - (performance bond) – a form of guarantee issued by a bank or insurance company to insure the client, up to a specified value, should the contractor fail to fulfil their obligations as detailed in the contract.
Change order - (variation order) – the written agreement between parties, setting out the costs and scope of additional work, or change to the contract.
Claim – a demand from one of the contracting parties for adjustment to the contract.
Civil construction – construction of concrete structures, roads or railways.
Client – the party who employed and contracted with the contractor. The client may be the owner of the facility, the managing contractor, or another contractor. Normally the client is the party that pays the contractor.
Client deliverables – the information and documentation the client requires from the contractor.
Commissioning – the process of testing the equipment and systems installed as part of the construction process.
Concurrent activities – activities which happen at the same time.
Construction – the physical work of building or constructing a facility (building, structure, road, dam or factory).
Contract – the agreement between the client and contractor.
Contract Administrators - (quantity surveyor) – a person who looks after the contractor's project finances, through preparing valuations, claims, cost reports and paying subcontractors.
Contract amendment – specific paperwork used during the construction process should anything change from the original contract. These changes could be additional work, the omission of work, changes to specifications or the project duration.
Contract document – the document which form the basis of the contract between the parties. They include drawings, terms and conditions, specifications and sets out the requirements for constructing the project.

Contract schedule - (contract program or programme) – the schedule which the client has agreed is the official one for the contract, it's used to measure progress, adjudicate any extension of time claims, and if necessary, to quantify the amount of the liquidated damages.
Contractor – a company that constructs or builds a facility, or a portion of the facility, for a client.
Cost-plus - (cost-plus a fee) – when the contractor is reimbursed their actual costs incurred in carrying out the contract, or variation, as well as a mark-up on these costs which is proportional to the costs and is normally expressed as a percentage.
Cost-recovery – the same as cost-plus.
Critical path – a sequence of activities linked together whose delay will affect the overall project completion.
Day-works – similar to cost-plus. Normally the contractor would specify rates for items of equipment and different trades people in the tender document, and these day-works rates are then used to calculate the cost of any additional work which the client may request the contractor to do, which cannot be costed out using the standard tendered unit rates. Often these rates would include a percentage for supervision, overheads and profit.
Demobilisation – the process of moving off site when the project is complete.
Designer – Architect or Engineer that designs the structures and facilities.
Design and construct contracts – when the contractor is responsible for both the design and the construction of the facility.
Design, build and finance contract – when the contractor is contracted to both design and construct the facility and, in addition, they must provide the finance for the project for the construction period and sometimes a set period beyond completion. When the client finally pays the contractor this payment would include the finance costs.
Design, build and operate contract – when the contractor is contracted to design and build the facility and then, after construction is complete, to operate the facility for a set period of time or for its life. The contractor may be paid for the period they operate the facility, a rental, or per unit of output from the facility.
Design indemnity insurance – insurance which the Designer should have in place in the event that their design is flawed, the structure requires repair or has to be replaced. In many instances the value of this insurance should cover the replacement of the structure.
Direct costs – the costs which can be directly attributable to carrying out a particular task, these include labour, material and equipment costs.
Drawings (plans) – graphic representation of the structures and facilities.
Employment contract – the contract between the employer and the employee which defines the conditions of employment.
Engineer – I have in most cases used this specifically to denote the engineer employed by the contractor to assist with managing the works.
Estimator – the contractor's person who prepares the tender or estimate.
Exclusions – items which the contractor may have excluded from their tender price.

External hire – equipment that is hired, or rented from another company, or external supplier.

Final account – the final value of the completed work, which includes the original contract value plus all contract amendments and back-charges. The contractor would have a final account with their client, and the contractor should have a final account with each of their subcontractors.

Float – the amount of time that a task can be delayed without impacting the final project completion date.

Force majeure – unforeseeable course of events which none of the contracting parties has any ability to prevent.

Foreman (Supervisor) – the person responsible to supervise the workers or a section of works.

Formwork (shutters) – the forms or structures used to shape and contain the wet concrete used in structures until it has gained sufficient strength to support itself.

Form-workers – the personnel that erect and remove the formwork.

Guarantees – a promise or assurance that an obligation will be met.

General conditions – the terms and conditions that apply to the work as a whole and which are often used for other projects.

Hold points – a specified point in the quality control procedure which has to be completed before the next step of the process can begin. A hold point may for instance be where the client has to inspect the work before further work is done.

Insurances – cover for potential losses. A risk management tool used to hedge against possible losses.

Information schedule – a schedule, or list, of when the client must make information available so as not to delay the contractor.

Job hazard assessments – an assessment made of a task or activity to estimate its risks, and what precautions and mitigating actions can be taken to minimise the risk and to lessen the impact should the risk eventuate.

Joint ventures – when two or more contractors enter into an agreement to jointly tender for and contract to a client to construct a facility and in doing so to share resources and risks.

Laydown area – the designated area on a construction site where the contractor can establish their facilities, and store their equipment and materials.

Leading-hand – a worker who is designated to supervise a small team of workers.

Lead time – the amount of time taken for an item to be delivered to the project. This time includes the time to design, manufacture and transport it to site.

Letter of intent – a letter issued by the client to the contractor informing them that it is their intention to award them the contract. It's normally treated as an instruction to start the works and the contract document will follow.

Lien – in certain contracts and countries when a contractor hasn't been paid by the client the contractor has the right to take occupation of the facility, even forcing its sale, so that they can recover the money the client owes them.

Liquidated damages – a specified amount of money which the contractor will pay the client should the contractor fail to meet the agreed contract completion dates.

Litigation – the process of using the court system to resolve a dispute.
Lost time injury – an injury, incurred on site, which results in the injured person being unable to return to work immediately after receiving treatment, or on the next scheduled work-day.
Managing contractor – the contractor appointed by the client to manage the project. This could also include a specialist project management company appointed to look after the client's or owner's interests and to manage the design team and the contractor.
Mark-up – the profit margin, although in some cases it may include the profit plus the contractor's overheads.
Materials – all items permanently incorporated by the contractor into the works. This may include concrete, reinforcing, road materials, building products, and including specialist items of equipment.
Mechanical work – the construction trades, which includes supplying and installing structural steel, mechanical equipment and piping.
Milestone – an important event, such as granting access or a completion date.
Mock-up – a model or small sample built to evaluate details and quality of the final item.
Monthly valuation – an assessment of the work that the contractor has completed during the month which reflects how much the client should pay.
Negotiate – to try and reach an equitable agreement through discussion.
Nominated subcontractor – a subcontractor the client specifies the contractor must use for a particular portion of the work.
Operators – personnel that drive or operate a piece of equipment, a vehicle or machine, but excluding the use of small hand tools.
Overhead costs - (indirect costs) – overhead project costs are costs the contractor incurs to run the project which cannot be directly related to specific tasks, such as the provision of management, supervision, site facilities, insurances and bonds. Company overheads are the costs a company incurs which are not directly attributable to a specific project, but are related to running the company and include costs such as Head Office rental, management and various support departments, such as finance and tendering.
Performance bond – an insurance issued by a bank, or insurance company, to the client that the contractor will complete the works to the required specifications.
Personal protective equipment – equipment issued to personnel for protection at work, this would include safety boots, helmets, gloves, safety glasses and overalls.
Planner - (Programmer, or Scheduler) – the person whose specific task is to prepare and update schedules.
Plant – a construction machine (such as an excavator, crane, dozer, loader, and including trucks and vehicles) used on a construction site to perform the work.
Plant and equipment – any item of equipment required to carry out the construction work, but not incorporated into the facility. It includes construction machines, hand tools and formwork.

Prestart inspections – an inspection carried out on a piece of plant or equipment to assess that it's in a fit state to use.

Project – any construction work.

Project Director – a person who is responsible to manage a number of construction projects and who may have several Project Managers reporting to them.

Project labour agreement – a specific labour agreement established for a project site, which governs the employment conditions (such as hourly rates, working hours, allowances, and project rules), which apply to all workers employed on the project.

Project Manager - (Site Manager, Construction Manager or Site Agent) – the person responsible to manage the contractor's work on the construction project.

Project organisation chart – a chart which shows the staff employed on a construction project in a diagrammatical format, and indicates their job title, responsibilities and their interrelationship and reporting structures.

Project safety plan – the overall safety plan detailing all the tasks and risks of carrying out these tasks, as well as the mitigating measures to reduce the risks and lessen the impact should any of the risks eventuate.

Punch lists - (snag list or defects list) – a list of outstanding items or repairs that must be completed so that the facility complies with the client's requirements.

Quality – the properties of the product supplied to the client, defined by the requirements in the contract document, which may include the visual appearance, as well as, the strength and durability.

Quality control – measures and procedures to ensure the product provided to the client meets the required quality.

Quality Control Managers – the contractor's representative responsible for ensuring that the quality control systems and procedures are in place and implemented.

Quality plan – the plan drawn up for the contractor to follow to ensure the work meets the required standards and specifications, and to monitor, track and report the procedures implemented by the contractor to ensure the work meets the required quality.

Quantity Surveyors – individuals who calculate the quantities on the construction project. They may also be responsible for project cost reports, claims, valuations and accessing and paying subcontractor's claims.

Register – any list, or log, maintained on the construction site to keep track of items, documents or inspections.

Reinforcing – the steel bars incorporated into concrete structures to give added strength.

Reinforcing-hands – personnel that place and fix (or tie), reinforcing in position.

Re-measurable contract – a contract where at the end of the project all of the work completed by the contractor is measured, and providing it meets the specifications, is paid for by the client.

Retention – a portion of money that is owed to the contractor but is withheld by the client, as insurance, until the contractor has fulfilled all their contractual obligations.

Safety Advisors – personnel employed by the contractor to assist with managing the safety procedures on a construction project to ensure it complies with the standards and requirements.

Safety Representatives – construction personnel, normally workers elected by fellow workers, who are part of a safety committee with the responsibility of ensuring that the construction works are undertaken in a safe manner.

Scaffolder – personnel used to erect scaffolding.

Scaffolding – temporary structures and platforms to enable workers to reach an elevated work area.

Schedule - (often referred to as a programme, program, bar chart or Gantt chart) – a graphic representation of the timetable needed to complete the project, showing the sequencing and duration of the various project tasks and activities.

Schedule link – the relationship between the activities on the schedule.

Scope of works – the work which the contractor is contracted to do. The scope normally takes the form of a written description of the work contained within the contract document.

Section Engineers – an engineer delegated to look after a section of works.

Self-perform – when the contractor does the work using their own employees rather than using subcontractors.

Shop drawings – drawings produced (normally by the contractor, their subcontractors or suppliers) to show the details of an item they have to fabricate.

Silo – a vertical steel or concrete storage structure, cylindrical in shape.

Site - (project site) – the area where the final construction of the facility takes place.

Site Administrators – personnel responsible for the contractor's general site administration work, such as preparing time sheets.

Site establishment – temporary facilities that the contractor builds to enable them to carry out the construction work.

Site facilities – the contractor's temporary buildings which include offices, workshops, toilets, eating areas, store buildings, and so on.

Slip-form shutters – concrete forms which are lifted on a continuous basis while the concrete is poured into them.

Specifications – definitions of the materials, processes and the quality products and systems.

Staff – the contractor's management, supervisory and support personnel who are generally paid a salary, usually personnel who aren't considered to be workers.

Standards – regulatory codes.

Subcontractor – a contractor employed by a contractor to do a portion of their works. The subcontractor would employ the personnel to do the work.

Substantial completion – when the whole or a section of the work can be occupied by the client.

Supervisors - (foreman) – the person who supervisors the contractor's workers or a section of works.

Support-work – scaffolding built to support formwork, particularly elevated formwork, (such as under slabs and beams), until the concrete reaches its design strength.

Sureties – a form of insurance supplied by a bank or insurance company to ensure that the contractor complies with their contractual obligations.

Survey – to set out the position of structures to be built, or to accurately work out the height and location of existing structures.

Tender - (bid, estimate or quote) – a price, or quotation, to carry out work submitted by the contractor to the client.

Tender documents - (bid documents or quotation) – the documents the client sends to the contractor so they can price the construction of the project. These documents would typically include details of the project, scope of works, specifications and drawings, and sufficient information for the contractor to price the work.

Tender submission - (bid submission) – the contractor's response to the tender document, it would include their price, as well as all supporting documentation and any other information, which the client required as part of the tender

Third party liability insurance – insurance taken out by the client or the contractor in case that the contractor damages property, or injures people, that are unrelated to them or their construction activities.

Toolbox meeting – meetings normally held weekly (or more frequently) with employees to discuss safety and any other items which may affect their work.

Tower crane – a crane consisting of a tower with a jib or arm on the top, which is normally used to construct high-rise buildings and structures.

Union - (trade union) – a body that represents the worker's rights

Variation a change from the original agreed contract.

Warranty – a guarantee that the product will function as it should.

Worker – manual and industrial or trades people who are generally employed on hourly or daily wages, and who physically do the work.

Worker's compensation insurance – insurance which covers the treatment of personnel injured on the project, as well as paying a portion of their wages which they can't earn while they are recuperating. The insurance may also pay their wages should they become permanently incapacitated and unable to work again due to the accident.

References

Dykstra, Alison. *Construction Project Management: A Complete Introduction*, Kirshner Publishing Company, INC.

Walker, Anthony. *Project Management in Construction: 5th Edition*, Blackwell Publishing

Mincks, William R & Johnston, Hal. *Construction Jobsite Management: 2nd Edition*, Thomson Delmar Learning

Jackson, Barbara J. *Construction Management Jumpstart: 2nd Edition*, Sybex an Imprint of Wiley

Civitello, Jr. Andrew M & Levy, Sidney M. *Construction Operations Manual of Policies and Procedures: 4th Edition*, McGraw-Hill

Halpin, Daniel W & Senior, Bolivar A. *Construction Management: 4th Edition*, Hamilton Printing

Schexnayder, Clifford J & Mayo, Richard E. *Construction Management Fundamentals*, McGraw-Hill

Made in the USA
Las Vegas, NV
09 November 2024

11452109R00142